FIREFIGHTER FATALITIES IN THE UNITED STATES IN 2007

Prepared for

United States Department of Homeland Security

Federal Emergency Management Agency

U.S. Fire Administration

National Fire Data Center

Contract Number EME-2003-CO-0282

Prepared by

C^2 Technologies, Inc.

Vienna, Virginia

June 2008

In memory of all firefighters
who answered their last call in 2007

To their families and friends

To their service and sacrifice

U.S. Fire Administration
Mission Statement

We provide National leadership to foster a solid foundation for local fire and emergency services for prevention, preparedness, and response.

TABLE OF CONTENTS

continued on next page

ACKNOWLEDGMENTS

This study of firefighter fatalities would not have been possible without the cooperation and assistance of many members of the fire service across the United States. Members of individual fire departments, chief fire officers, wildland fire service organizations such as the United States Forest Service, the National Park Service, the Bureau of Land Management, the Bureau of Indian Affairs, the United States Fish and Wildlife Service, as well as the United States Department of Justice, the National Fire Protection Association (NFPA), the National Fallen Firefighters Foundation (NFFF), and many others contributed important information for this report.

C^2 Technologies, Inc., of Vienna, Virginia, conducted this analysis for the U.S. Fire Administration (USFA) under contract EME-2003-CO-0282.

The ultimate objective of this effort is to reduce the number of firefighter deaths through an increased awareness and understanding of their causes and how they can be prevented. Firefighting, rescue, and other types of emergency operations are essential activities in an inherently dangerous profession, and unfortunate tragedies do occur. This is the risk all firefighters accept every time they respond to an emergency incident. However, the risk can be greatly reduced through efforts to improve training, emergency scene operations, and firefighter health and safety initiatives.

Cover photo courtesy of Glen E. Ellman, www.fortworthfire.com

BACKGROUND

For 31 years, the USFA has tracked the number of firefighter fatalities and conducted an annual analysis. Through the collection of information on the causes of firefighter deaths, the USFA is able to focus on specific problems and direct efforts toward finding solutions to reduce the number of firefighter fatalities in the future. This information also is used to measure the effectiveness of current programs directed toward firefighter health and safety.

Several programs have been funded by the USFA in response to this detailing of firefighter fatalities. For example, the USFA has sponsored significant work in the areas of general emergency vehicle operation safety, fire department tanker/tender operation safety, firefighter incident scene rehabilitation, and roadside incident safety. The data developed for this report also are used widely in other firefighter fatality prevention efforts.

One of the USFA's main program goals is a 25-percent reduction in firefighter fatalities in 5 years, and a 50-percent reduction within 10 years. The emphasis placed on these goals by the USFA is underscored by the fact that these goals represent one of the five major objectives that guide the actions of the USFA.

In addition to the analysis, the USFA provides a list of firefighter fatalities and associated documentation to the NFFF. If certain criteria are met, the fallen firefighter's next of kin, as well as members of the individual's fire department, are invited to the annual Fallen Firefighters Memorial Service. The service is held at the National Emergency Training Center (NETC) in Emmitsburg, Maryland, during Fire Prevention Week in October of each year. Additional information regarding the Memorial Service can be found at www.firehero.org or by calling the NFFF at (301) 447-1365.

Other resources and information regarding firefighter fatalities, including current fatality notices, the National Fallen Firefighters Memorial database, and links to the Public Safety Officers' Benefit (PSOB) Program can be found at www.usfa.dhs.gov/fireservice/fatalities/

INTRODUCTION

This report continues a series of annual studies by the USFA of onduty firefighter fatalities in the United States.

The specific objective of this study is to identify all onduty firefighter fatalities that occurred in the United States and its protectorates in 2007, and to analyze the circumstances surrounding each occurrence. The study is intended to help identify approaches that could reduce the number of firefighter deaths in future years.

In addition to the 2007 overall findings, this study includes information on seatbelt use for firefighters and efforts to encourage seatbelt use.

WHO IS A FIREFIGHTER?

For the purpose of this study, the term "firefighter" covers all members of organized fire departments in all 50 States, the District of Columbia, and the Territories of Puerto Rico, the Virgin Islands, American Samoa, the Commonwealth of the Northern Mariana Islands, and Guam. It includes career and volunteer firefighters; full-time public Safety Officers acting as firefighters; State, Territory, and Federal government fire service personnel, including wildland firefighters; and privately employed firefighters, including employees of contract fire departments and trained members of industrial fire brigades, whether full- or part-time. It also includes contract personnel working as firefighters or assigned to work in direct support of fire service organizations.

Under this definition, the study includes not only local and municipal firefighters but also seasonal and full-time employees of the United States Forest Service, the Bureau of Land Management, the Bureau of Indian Affairs, the U.S. Fish and Wildlife Service, the National Park Service, and State wildland agencies. The definition also includes prison inmates serving on firefighting crews; firefighters employed by other governmental agencies, such as the U.S. Department of Energy (DOE); military personnel performing assigned fire suppression activities; and civilian firefighters working at military installations.

WHAT CONSTITUTES AN ONDUTY FATALITY?

Onduty fatalities include any injury or illness sustained while on duty that proves fatal. The term "onduty" refers to being involved in operations at the scene of an emergency, whether it is a fire or nonfire incident; responding to or returning from an incident; performing other officially assigned duties such as training, maintenance, public education, inspection, investigations, court testimony, and fundraising; and being on call, under orders, or on standby duty except at the individual's home or place of business. An individual who experiences a heart attack or other fatal injury at home while he or she prepares to respond to an emergency is considered on duty when the response begins. A firefighter who becomes ill while performing fire department duties and suffers a heart attack shortly after arriving home or at another location may be considered on duty since the inception of the heart attack occurred while the firefighter was on duty.

On December 15, 2003, the President of the United States signed into law the Hometown Heroes Survivors Benefit Act of 2003. After being signed by the President, the Act became Public Law 108-182. The law presumes that a heart attack or stroke are in the line of duty if the firefighter was engaged in nonroutine stressful or strenuous physical activity while on duty and the firefighter becomes ill while on duty or within 24 hours after engaging in such activity. The full text of the law is available at: http://frwebgate.access. gpo.gov/cgi-bin/getdoc.cgi?dbname=108_cong_ public_laws&docid=f:publ182.108.pdf

The inclusion criteria for this study have been affected by this change in the law. Previous to December 15, 2003, firefighters who became ill as the result of a heart attack or stroke after going off duty needed to register some complaint of not feeling well while still on duty in order to be included in this study. For firefighter fatalities after December 15, 2003, firefighters will be included in this study if they become ill as the result of a heart attack or stroke within 24 hours of a training activity or emergency response. Firefighters who become ill after going off duty where the activities while on duty were limited to tasks that did not involve physical or mental stress will not be included in this study.

A fatality may be caused directly by an accidental or intentional injury in either emergency or nonemergency circumstances, or it may be attributed to an occupationally related fatal illness. A common example of a fatal illness incurred on duty is a heart attack. Fatalities attributed to occupational illnesses would also include a communicable disease contracted while on duty that proved fatal when the disease could be attributed to a documented occupational exposure.

Firefighter fatalities are included in this report even when death is considerably delayed after the original incident. When the incident and the death occur in different years, the analysis counts the fatality as having occurred in the year in which the incident took place. One firefighter died in 2006 whose death was not known to the USFA until 2007. Information about this death is included in Appendix A of this report, but the death is not addressed in the body of the report unless it affects retrospective statistical comparisons.

There is no established mechanism for identifying fatalities that result from illnesses such as cancer that develop over long periods of time and which may be related to occupational exposure to hazardous materials or toxic products of combustion. It has proved to be very difficult over the years to provide a complete evaluation of an occupational illness as a causal factor in firefighter deaths due to the following limitations: the exposure of firefighters to toxic hazards is not sufficiently tracked; the often-delayed long-term effects of such toxic hazard exposures; and the exposures firefighters may receive while off duty.

SOURCES OF INITIAL NOTIFICATION

As an integral part of its ongoing program to collect and analyze fire data, USFA solicits information on firefighter fatalities directly from the fire service and from a wide range of other sources. These sources include the PSOB Program administered by the U.S. Department of Justice (DOJ), the National Institute for Occupational Safety and Health (NIOSH), the Occupational Safety and Health Administration (OSHA), the Department of Defense (DOD), the National Interagency Fire Center, and other Federal agencies.

The USFA receives notification of some deaths directly from fire departments, as well as from such fire service organizations as the International Association of Fire Chiefs (IAFC), the International Association of Fire Fighters (IAFF), the National Fire Protection Association (NFPA), the National Volunteer Fire Council (NVFC), State fire marshals, State fire training organizations, other State and local organizations, fire service Internet sites, news services, and fire service publications. The USFA also keeps track of fatal fire incidents as part of its Major Fires Investigation Program and performs an ongoing analysis of data from the National Fire Incident Reporting System (NFIRS).

PROCEDURE FOR INCLUDING A FATALITY IN THE STUDY

In most cases, after notification of a fatal incident, initial telephone contact is made with local authorities by the USFA to verify the incident, its location, jurisdiction, and the fire department or agency involved. Further information about the deceased firefighter and the incident may be obtained from the chief of the fire department or designee over the phone or by other data collection forms. After basic information is collected, a notice of the firefighter fatality is posted at the USFA site in Emmitsburg, Maryland, and a notice of the fatality is transmitted by electronic mail to a large list of fire service organizations and fire service members.

Information that is requested routinely from fire departments that have experienced a fatality includes NFIRS-1 (incident) and NFIRS-3 (fire service casualty) reports, the fire department's own incident and internal investigation reports, copies of death certificates and autopsy results, special investigative reports, law enforcement reports, photographs and diagrams, and newspaper or media accounts of the incident. Information on the incident also may be gathered from NFPA or NIOSH reports on an incident.

After obtaining this information, a determination is made as to whether the death qualifies as an onduty firefighter fatality according to the previously described criteria. With the exception of firefighter deaths after December 15, 2003, the same criteria were used for this study as in previous annual studies. Additional information may be requested by USFA, either through follow up with the fire department directly, from State vital records offices, or from other agencies. The final determination as to whether a fatality qualifies as an onduty death for inclusion in this statistical analysis is made by the USFA. The final determination as to whether a fatality qualifies as a line-of-duty death for inclusion in the annual Fallen Firefighters Memorial Service is made by the NFFF.

Photo by Mark Whitney, U.S. Fire Administration

2007 FINDINGS

One hundred and eighteen (118) firefighters died while on duty in 2007. This total includes 13 firefighters who are included in this report as a result of the inclusion criteria changes resulting from the Hometown Heroes Act of 2003.

After a 2-year respite, the number of firefighter fatalities using the pre-Hometown Heroes criteria was above 100. Using the pre-Hometown Heroes criteria, 105 firefighters died while on duty in the United States in 2007, up dramatically from the 92 onduty firefighter deaths in 2006 using pre-Hometown Heroes inclusion criteria. The lowest number of onduty firefighter fatalities recorded within the past three decades was 1992 with 77 fatalities, followed by 1993 with 81 fatalities (Figure 1).

In December of 2003, the Hometown Heroes Survivors Benefit Act (the Act) of 2003 was signed into law. For Federal survivor's benefit purposes, the law presumes that a heart attack or stroke are in the line of duty if the firefighter was engaged in nonroutine stressful or strenuous physical activity while on duty and the firefighter becomes ill while on duty or within 24 hours after engaging in such activity. Prior to this law, Federal survivor's benefits for firefighters generally were not paid for heart attacks or strokes, regardless of the circumstances.

The inclusion criteria for this report also were modified when the Act became law. Prior to December of 2003, a firefighter needed to express or show signs of illness prior to going off duty

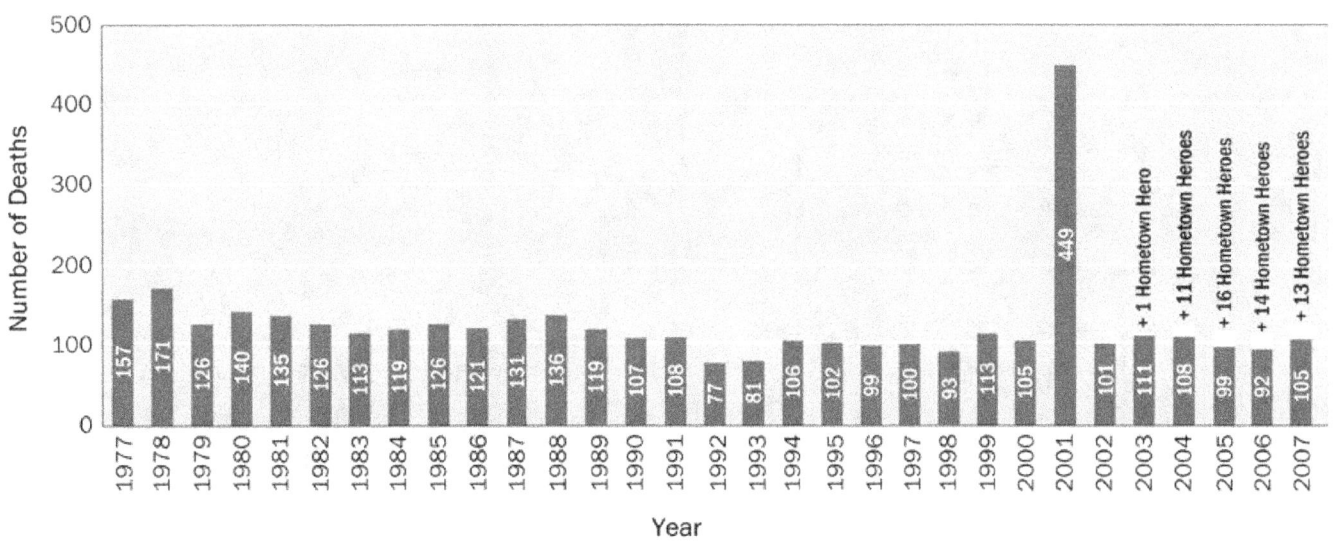

Figure 1. Onduty Firefighter Fatalities (1977-2007)

in order to be included in this report. After December of 2003, firefighters who became ill within 24 hours of onduty stressful or strenuous activity also were included.

While a change in reporting criteria does not diminish the sacrifices made by the firefighters who die, or the sacrifices made by their families and their peers, statistical analysis of death trends needs to acknowledge the changes stemming from the Act.

In 2001, the largest loss of firefighters' lives in a single incident in history occurred as a result of the attacks on the World Trade Center (WTC) in New York City on September 11th. A total of 344 firefighters were killed in the attacks and resulting collapses. When conducting multiyear comparisons of firefighter fatalities in this report, it may be necessary to set these deaths apart for illustrative purposes. This action is by no means a minimization of the supreme sacrifice made by these firefighters.

CAREER AND VOLUNTEER DEATHS

In 2007, onduty firefighter fatalities included 68 volunteer firefighters and 50 career firefighters (Figure 2). Among the volunteer firefighter fatalities, 63 were from local or municipal volunteer fire departments, and 5 were part-time or full-time members of wildland fire agencies. One career firefighter was a member of an industrial fire department, and the rest were members of local or municipal fire departments.

> **The number of career firefighter deaths rose significantly from 29 in 2006 to 50 in 2007.**

Three of the firefighters who died in 2007 were female and 115 were male. In the last decade, female firefighter deaths have ranged from a low of 0 in 1998 to a high of 6 in 2004 and in 2006.

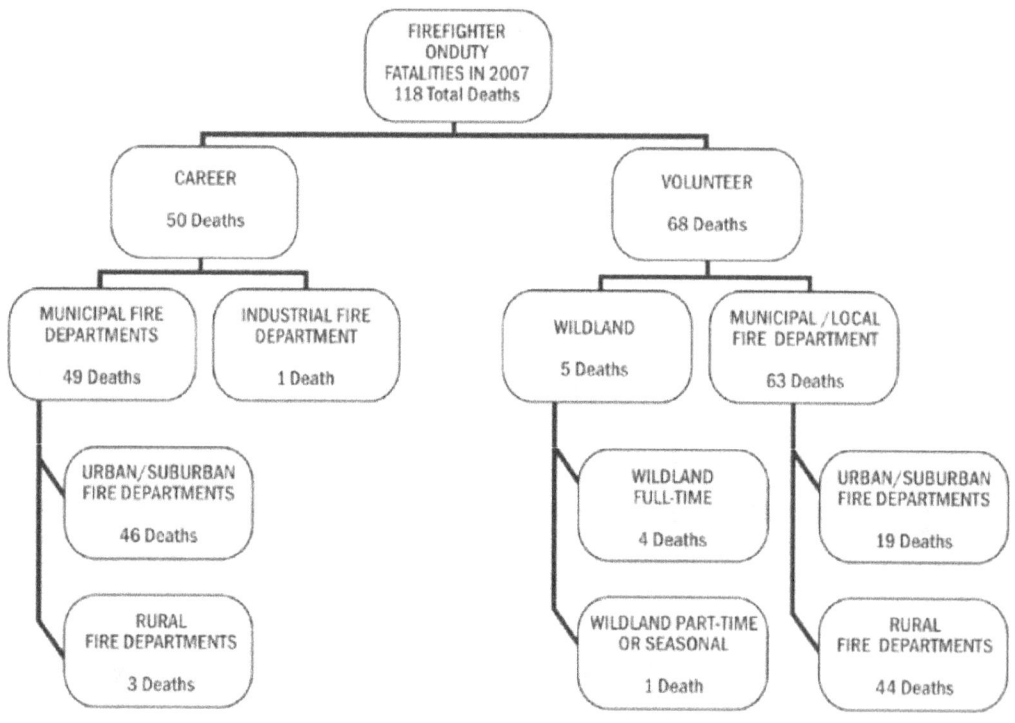

Figure 2. Career and Volunteer Deaths (2007).

MULTIPLE FIREFIGHTER FATALITY INCIDENTS

The 118 deaths in 2007 resulted from a total of 104 fatal incidents. There were 7 firefighter fatality incidents where 2 or more firefighters were killed in 2007, claiming a total of 21 firefighters' lives.

- On January 30th, two West Virginia firefighters were killed when a propane tank leak at a convenience store caused an explosion.

- On March 24th, two North Carolina firefighters were killed in the crash of a water tanker (tender) while responding to a reported structure fire.

- On June 18th, nine Charleston, South Carolina, firefighters were killed when they became disoriented in a fast-moving fire in a retail furniture store.

- On July 21st, two California firefighters died when they became trapped by rapid fire progress in a residential fire.

- On August 3rd, two Texas firefighters died during firefighting efforts in a residence.

Table 1: Multiple Firefighter Fatality Incidents

Year	Number of Incidents	Total Number of Deaths
2007	7	21
2006	6	17
2005	4	10
2004	3	6
2003	7	20
2002	9	25
2001	8	362
2001 w/o WTC	7	18
2000	5	10
1999	6	22
1998	10	22

On June 18, 2007, nine South Carolina firefighters died in a structure fire. Exclusive of the 9/11 attack, this is the highest loss of firefighter's lives in a structure fire since the Hotel Vendome collapse in Boston on June 17, 1972 that claimed the lives of nine firefighters.

- On August 18th, two New York City firefighters died in a highrise fire that occurred in a building undergoing demolition as a result of damage the building received in the attacks of September 11, 2001.

- On August 29th, two Boston firefighters died when a fire in a small restaurant suddenly expanded and trapped them.

WILDLAND FIREFIGHTING DEATHS

In 2007, 11 firefighters were killed during activities involving brush, grass, or wildland firefighting. This total includes part-time and seasonal wildland firefighters, full-time wildland firefighters, and municipal or volunteer firefighters whose deaths are related to a wildland fire (Figure 3). This is the lowest level of wildland-related firefighter deaths in over a decade.

Five firefighters were killed in vehicle crashes related to wildland firefighting duties. Two of those firefighters were killed in crashes involving their personal vehicles as they responded to wildland incidents; one in Indiana and one in Pennsylvania. A South Carolina firefighter died in a crash as he returned to his base station after having participated in a day-long prescribed burn, and a North Carolina firefighter was killed when a crash on the opposite side of a divided highway caused a tractor-trailer truck to cross over the median into the firefighter's lane. A California firefighter was killed when the bulldozer he was operating overturned while working at a wildland fire.

Three firefighters engaged in wildland firefighting duties died as the result of heart attacks: An Ohio firefighter suffered a heart attack as he and other firefighters stood by in their fire station for a response to a mutual-aid wildland fire. An Idaho firefighter died in April after completing a wildland pack test and subsequently experiencing a heart attack. A Tennessee firefighter suffered a heart attack in August of 2007 as he flagged a proposed fire line.

A Kansas fire officer was killed when he came into contact with an energized powerline while fighting a wildland fire. The line had fallen to the ground when a power pole was struck by a truck, resulting in the fire.

In 2007, there were no multiple firefighter fatality incidents related to wildland firefighting. This is only the second year for no multiple wildland fatality incidents in over a decade.

A Florida firefighter was killed when he was crushed by a falling tree during power saw training in November.

The only aircraft firefighting death of 2007 occurred when a helicopter contacted trees during a supply dropoff operation. The helicopter crashed and the pilot was killed.

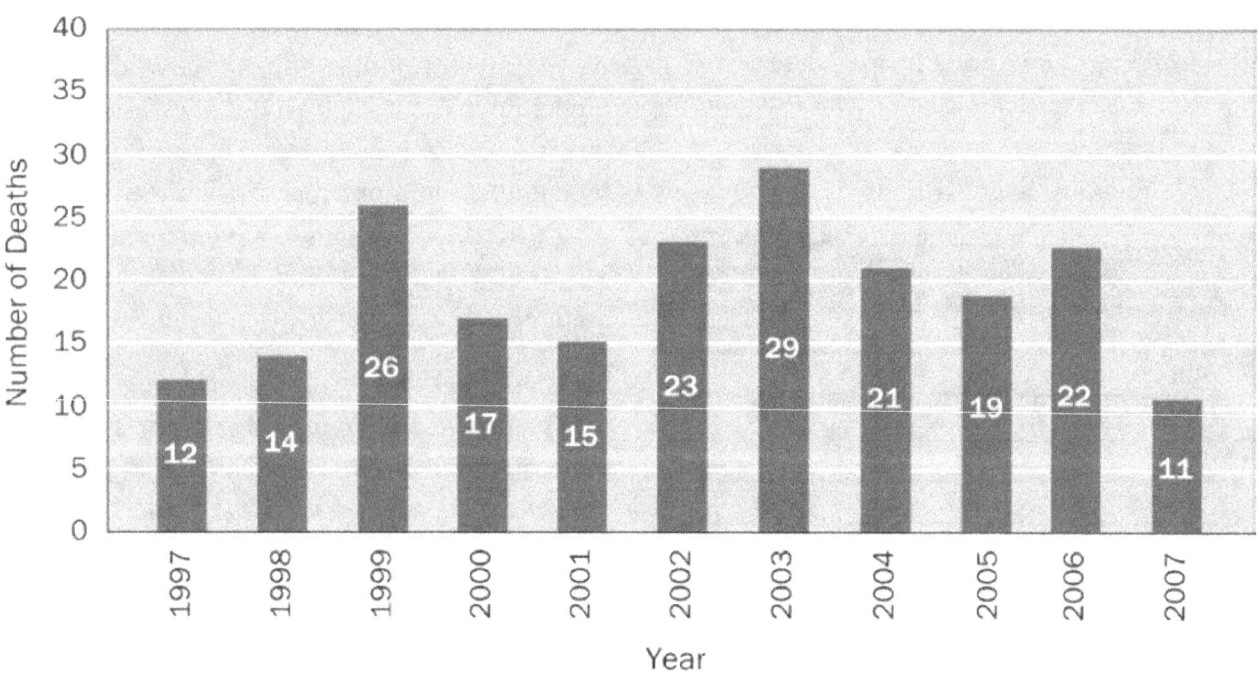

Figure 3. Firefighter Fatalities Related to Wildland Firefighting (1997-2007).

Table 3: Wildland Firefighting Aircraft Deaths

Year	Total Number of Deaths	Number of Fatal Incidents
2007	1	1
2006	8	3
2005	6	2
2004	3	3
2003	7	4
2002	6	3
2001	6	3
2000	6	5
1999	0	0
1998	3	2

Table 2: Firefighter Deaths Associated with Wildland Firefighting

Year	Total Number of Deaths	Number of Fatal Incidents	Number of Firefighters Killed in Multiple-Death Incidents
2007	11	11	0
2006	22	13	13
2005	19	15	6
2004	21	21	0
2003	30	22	10
2002	23	14	13
2001	15	9	9
2000	19	16	6
1999	27	26	2
1998	14	13	2

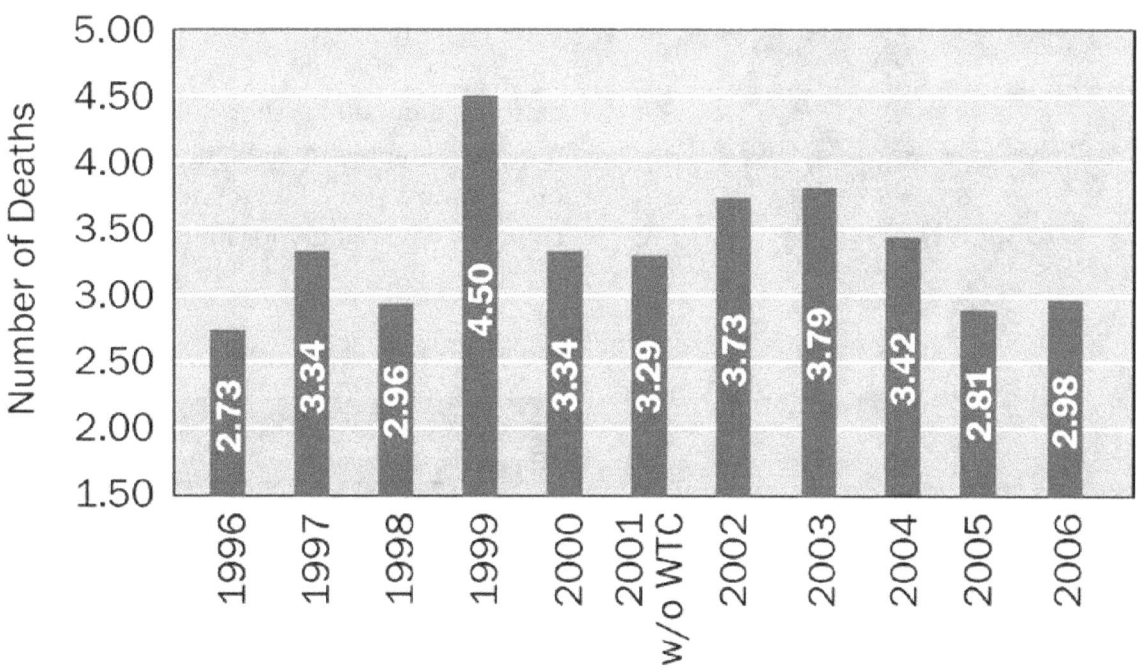

Figure 4. Firefighter Fatalities per 100,000 Fires.

TYPE OF DUTY

Activities related to emergency incidents resulted in the deaths of 76 firefighters in 2007, up from 61 such deaths in 2006 (Figure 5). This includes all firefighters who died while responding to an emergency, while at an emergency scene, while returning from the emergency incident, and other emergency-related activities. Nonemergency activities accounted for 42 fatalities. Nonemergency duties include training, administrative activities, performing other functions that are not related to an emergency incident, and postincident fatalities where the firefighter does not experience the illness or injury during the emergency.

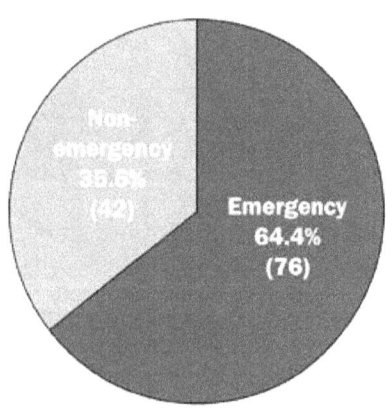

Figure 5. Firefighter Deaths by Type of Duty (2007).

A multiyear historical perspective concerning the percentage of firefighter deaths that occurred during emergency duty is presented in Table 4. The data for 2003 and after is skewed somewhat by the inclusion of firefighters covered by the changes resulting from the Hometown Heroes Act of 2003.

Table 4: Emergency Duty Firefighter Deaths

Year	Percentage of All Deaths	Percentage of All Deaths Without Hometown Heroes
2007	64.4	72.4
2006	57.5	66.3
2005	52.1	60.6
2004	68.9	75.9
2003	69	69.6
2002	73	n/a
2001	65	n/a
2001 with WTC	92	n/a
2000	71	n/a
1999	87	n/a
1998	77	n/a

The number of deaths by type of duty being performed in 2007 is shown in Table 5 and presented graphically in Figure 6. As has been the case for most years, fireground duties are the most common type of duty for firefighters killed while on duty.

Table 5: Firefighter Deaths by Type of Duty (2007)

Type of Duty	Number of Deaths
Fireground Operations	38
Responding/Returning	26
Other On Duty	20
Training	11
Nonfire Emergencies	8
After an Incident	15
Total	118

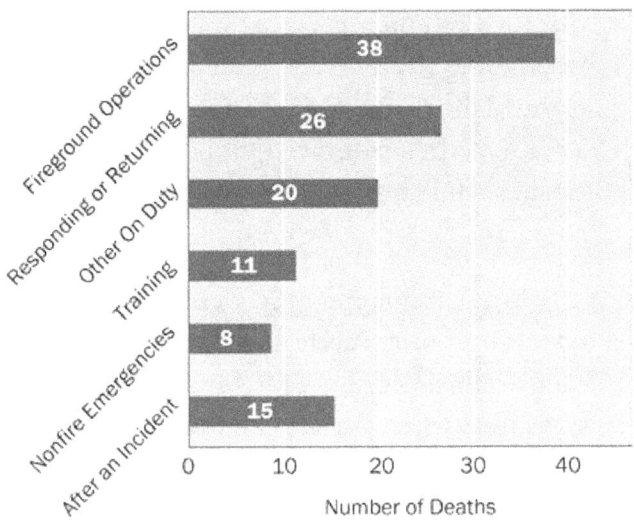

Figure 6. Fatalities by Type of Duty (2007).

FIREGROUND OPERATIONS

Thirty-eight firefighters died while engaged in activities at the scene of a fire in 2007. Traditionally the fire scene is the most hazardous work area for firefighters each year. When compared with the sheer number of responses to nonfire emergencies such as emergency medical incidents by firefighters, the fireground is more dangerous than the scene of a medical emergency by orders of magnitude. Thirty-five fireground deaths were associated with structure fires and three were associated with wildland fires.

Five of the seven multiple firefighter fatality incidents that occurred in 2007 were on the firegound. These incidents claimed the lives of 17 firefighters:

- Nine Charleston, South Carolina, firefighters were killed when they became disoriented and overcome by smoke and fire conditions during a fire in a retail furniture store.

- Two Contra Costa County, California, firefighters died when fire progressed rapidly during a residential structure fire.

- Two Tyler, Texas, area firefighters were killed in a residential structure fire in August.

- Two New York City firefighters were killed when they were unable to exit the hazardous area during a highrise fire in a building undergoing demolition.

- Two Boston firefighters died in a restaurant fire when fire conditions changed rapidly.

Nine firefighters suffered heart attacks at fire scenes in 2007, one more than in 2006. All nine heart attacks occurred at structure fires.

In addition to the multiple firefighter fatality incidents described above, 12 firefighters died of traumatic injuries on the fireground in 2007:

- Three firefighters were killed in separate incidents as a result of structural collapses. Two of the incidents occurred in Pennsylvania and involved motorcycles: The first involved the collapse of a former motorcycle garage that was on fire, and the other fire was caused by a gasoline spill from a motorcycle undergoing repair. The third incident occurred in Nebraska, and involved the collapse of a residence while firefighters were advancing hoselines to the interior.

- Two firefighters died in separate incidents when they fell into the basement of residences involved in fire. Both incidents occurred in January, one in Indiana and one in Tennessee. In one case, the structure used traditional full-dimension lumber and, in the other, engineered lumber products were involved.

- Two firefighters were killed in residential structure fires when they were trapped by rapid fire progress: one in Virginia and the other in Georgia.

- A Florida fire officer was injured at a structure fire and subsequently died of a pulmonary embolism when a blood clot from the injury travelled to his lungs.

- A New York City firefighter fell to his death from an aerial ladder (or from the roof near an aerial ladder) during a structure fire.

- A helicopter crash claimed the life of a pilot engaged in supply delivery duties at a wildland fire in California.

- A Kansas fire officer was electrocuted at a wildland fire.

- A California wildland firefighter was killed when he was crushed by the rollover of a bulldozer he was operating.

> **The failure of engineered lumber trusses under fire conditions has been a factor in at least 3 firefighter fatalities in 2006 and 2007.**

RESPONDING/RETURNING

Twenty-six firefighters died while responding to or returning from emergency incidents in 2007: 24 while responding to an emergency incident, and 2 while returning from an emergency.

> **Two of the three female firefighter fatalities in 2007 occurred in vehicle crashes while responding.**

Ten firefighters were killed in crashes that involved personal vehicles, nine while responding and one while returning from an emergency:

- Eight firefighters were killed in crashes that involved their personal vehicles while responding to an incident. In seven of the eight fatalities, the firefighter was not wearing a seatbelt. In six of the eight personal vehicle crashes while responding, excessive speed was citied as a factor in the crash.

- A North Carolina fire officer crashed while operating a motorcycle and returning from a motor vehicle crash. A Michigan firefighter was killed when a tree fell onto his personal vehicle as he responded to an incident.

Table 6: Firefighter Deaths While Responding to or Returning From an Incident

Year	Number of Firefighter Deaths
2007	26
2006	15
2005	22
2004	23
2003	36
2002	13
2001	23
2000	19
1999	26
1998	14

Six firefighters suffered heart attacks and one firefighter suffered a cerebrovascular accident (CVA) while responding to an incident or returning from an incident. Heart attacks or CVAs occurred as firefighters began their response from home, while at the fire station preparing to respond, while driving or as a passenger in an emergency vehicle, and while returning from an incident.

Four firefighters died as the result of tanker (tender) crashes while responding to an incident. Two North Carolina firefighters died in a tanker crash in March and two firefighters died in separate tanker crashes in Maine and Alabama; both crashes occurred in May.

Four firefighters died in crashes that involved fire apparatus: three in engines and one in a ladder truck. One firefighter died in a crash that involved a fire police van.

OTHER ON DUTY

In 2007, 20 firefighters died on duty while engaged in activities that were not on the scene of an emergency or associated with training:

- Fourteen firefighters suffered heart attacks while on duty but not assigned to an incident or emergency response.

- Two firefighters died as the result of injuries that they received after a fall from fire apparatus. One firefighter was loading hose and the other firefighter was loading an antique fire engine on a trailer when the falls occurred.

- In separate incidents, two California firefighters died in vehicle crashes while on duty.

- A Utah fire chief drowned after his fire department vehicle crashed into a reservoir.

- A South Carolina firefighter died in a vehicle crash as he drove back to base from a prescribed burn.

TRAINING

In 2007, 11 firefighters died while they were engaged in training activities:

- Seven of the deaths were heart attacks. Two firefighters experienced heart attacks while out of town for training events; two heart attacks occurred during or shortly after fitness evaluations; two heart attacks occurred during firefighting training; and one heart attack

occurred as the firefighter participated in physical fitness activities.

- Two firefighters were killed in vehicle crashes as they returned from mandatory training activities.

- A Maryland firefighter was killed when she was severely burned during live fire training in an acquired residential structure.

- A Florida firefighter was killed by a falling pine tree during wildland firefighter training.

Table 7 offers a multiyear perspective on training deaths.

Table 7: Firefighter Deaths During Training

Year	Number of Firefighter Deaths
2007	11
2006	9
2005	14
2004	13
2003	12
2002	11
2001	14
2000	13
1999	3
1998	12

NONFIRE EMERGENCIES

Eight firefighters died when they became ill or were injured while on the scene of emergencies that did not involve fire:

- Two West Virginia firefighters were killed when propane escaping from a storage cylinder at a local convenience store ignited and exploded.

- Two firefighters died at emergency medical incidents. Two more died of heart attacks at special operations incidents: one at a hazardous materials incident and one at a technical rescue.

- An Illinois firefighter was struck and killed by a passing bus while working on the scene of a disabled truck.

- An Arkansas firefighter fell from a bridge as he assisted at the scene of a motor vehicle crash.

AFTER THE INCIDENT

Fifteen firefighters died after the conclusion of their onduty activity. All 15 deaths were heart attacks. Twelve of the deaths were included in this study as a result of the Hometown Heroes Act inclusion criteria change.

CAREER, VOLUNTEER, AND WILDLAND DEATHS BY TYPE OF DUTY

Figure 7 depicts career, volunteer, and wildland firefighter deaths by type of duty. Wildland career, wildland seasonal, and wildland contractor deaths were grouped together. As in past years, a disproportionate number of fatalities was experienced by volunteer firefighters responding to and returning from alarms, as compared to career firefighters.

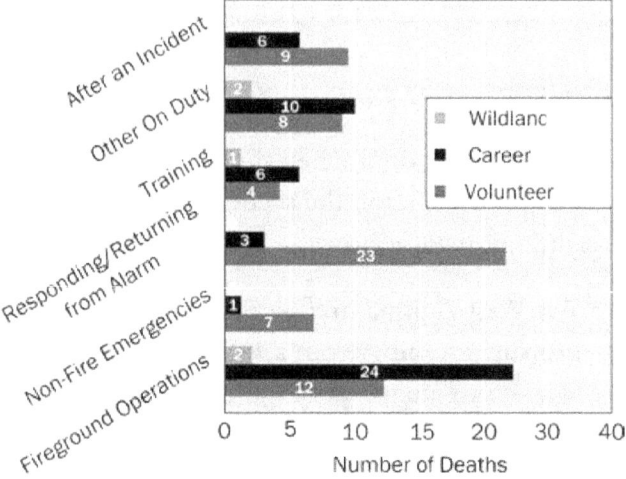

Figure 7. Career, Volunteer, and Wildland Deaths by Type of Duty (2007).

TYPE OF EMERGENCY DUTY

In 2007, 70 firefighters died while responding to or working on the scene of an emergency. This number includes deaths resulting from injuries sustained on the incident scene or en route to the incident scene and firefighters who became ill on an incident scene and later died. It does not include firefighters who became ill or died while returning from an incident (such as a vehicle collision while returning from an incident). Figure 8 shows the number of firefighters killed in firefighting, emergency medical services (EMS), hazardous materials-related incidents, and other emergency incidents in 2007.

Fifty-one firefighters were killed during firefighting duties; 12 firefighters were killed on EMS calls; 3 firefighters were killed in association with hazardous materials incidents; and 4 were killed in other emergency circumstances.

Note: 70 of 118 deaths occurred during emergency responses.

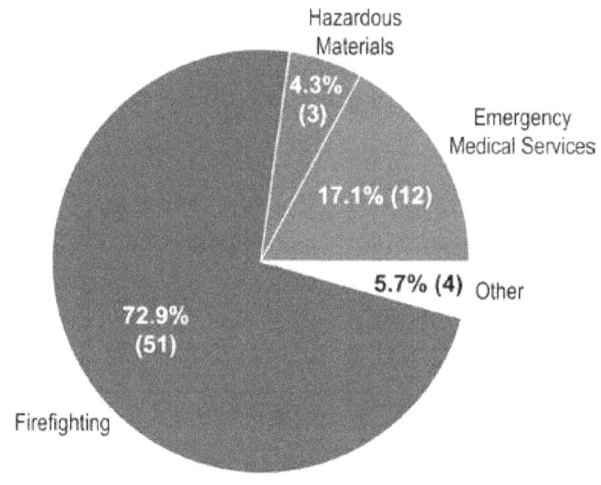

Figure 8. Type of Emergency Duty Including (2007).

CAUSE OF FATAL INJURY

The term "cause of injury" refers to the action, lack of action, or circumstances that resulted directly in the fatal injury. The term "nature of injury" refers to the medical cause of the fatal injury or illness; often this is referred to as the physiological cause of death. A fatal injury usually is the result of a chain of events, the first of which is recorded as the cause.

Table 8 and Figure 9 show the distribution of deaths by cause of fatal injury or illness.

Table 8: Cause of Fatal Injury (2007)

Cause	Number
Stress/Overexertion	55
Vehicle Collision	27
Lost/Disoriented	11
Caught/Trapped	7
Collapse	7
Struck by	5
Fall	4
Other	2
Total	118

STRESS OR OVEREXERTION

Stress or overexertion is a general category that includes all firefighter deaths that are cardiac or cerebrovascular in nature such as heart attacks and strokes (CVAs), and other events such as extreme climatic thermal exposure. Classification of a firefighter fatality in this cause of fatal injury category does not necessarily indicate that a firefighter was in poor physical condition.

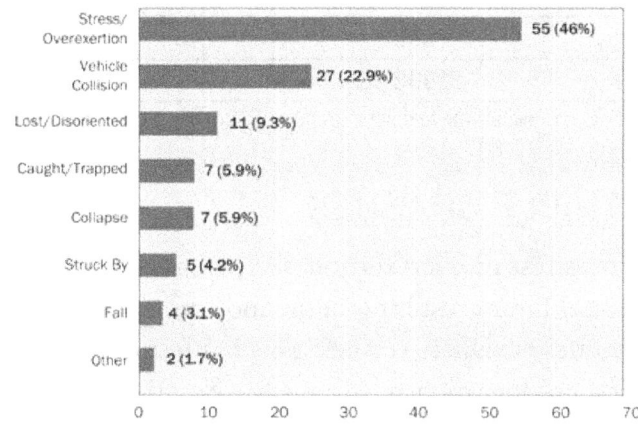

Figure 9. Fatalities by Cause of Fatal Injury (2007).
Note: Rounding error, does not equal 100%.

Firefighting is extremely strenuous physical work and is likely one of the most physically demanding activities that the human body performs.

- Fifty-five firefighters died in 2007 as a result of stress/overexertion:

  – Fifty-two of the stress deaths were heart attacks.

  – Two firefighters died due to a CVA.

  – One firefighter died of a cardiac condition that was not classified as a heart attack.

  – There were no extreme heat- or cold-related deaths in 2007.

- If the Hometown Heroes deaths in 2007 are set aside for analysis purposes, 40 percent of firefighter fatalities in 2007 were caused by stress or overexertion.

Table 9: Deaths Caused by Stress or Overexertion

Year	Number	Percent of Fatalities
2007	55	46.6
2006	54	50.9
2005	62	53.9
2004	66	56.4
2003	51	45.9
2002	38	38
2001	43	40.9*
2000	46	44.6
1999	56	49.5
1998	43	46.2

Does not include the firefighter deaths of September 11, 2001, in New York City.

VEHICLE CRASHES

After stress or overexertion, the perennial cause of fatal injury resulting in the most firefighter fatalities is vehicle crashes. This has been the case for a number of years. In many cases, these deaths appear to have been preventable (Figure 10).

- Twenty-seven firefighters were killed in 2007 as a result of vehicle crashes.

- One of these deaths occurred in an aircraft crash.

- Twenty-six firefighters were killed in nonaircraft vehicle crashes.

  – Eleven crashes involved the firefighter's personal vehicle.

  – Six crashes involved fire department vehicles that were not apparatus.

  – Three crashes and four deaths involved a fire department tanker (tender).

  – Four crashes involved other fire apparatus, three engines and a ladder.

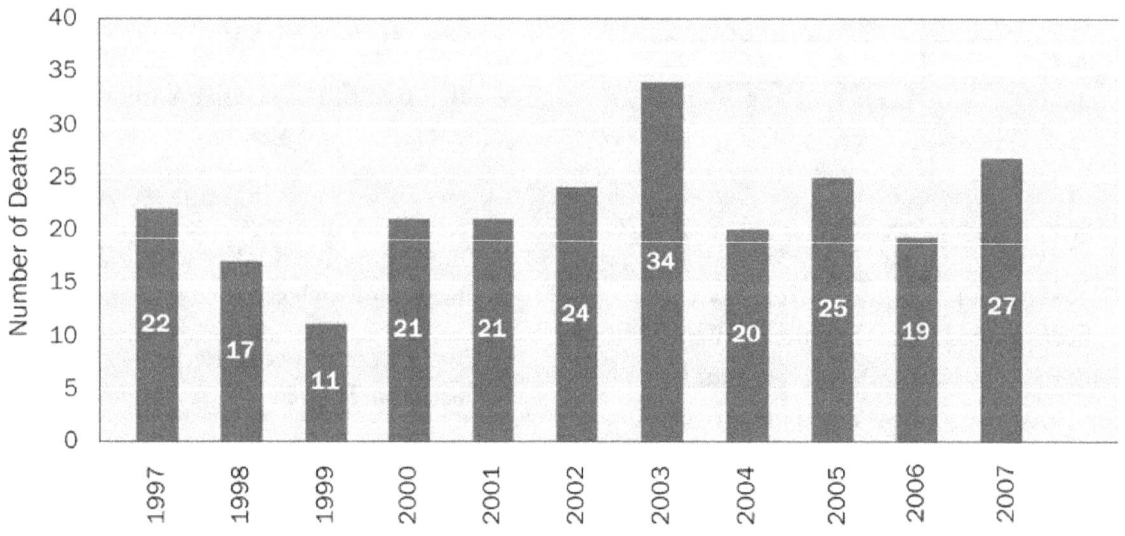

Figure 10. Firefighter Fatalities in Vehicle Collisions.

> *The 11 personal vehicle crash deaths in 2007 is the same number of firefighter deaths in personal vehicle crashes in 2005 and 2006 combined, though only slightly more than the 5-year average of 10 per year. Many of these deaths involved excessive speed and/or the lack of seatbelt use.*

– One crash involved a fire department bulldozer.

- No seatbelt was used in 11 of the 20 cases where seatbelts were available and the status of their use was known.

LOST OR DISORIENTED

In two multiple-fatality incidents, 11 firefighters died in 2007 when they became lost or disoriented inside of structure fires. Nine Charleston, South Carolina, firefighters died when they became disoriented in a large retail furniture store, and two New York City firefighters died in a fire involving a highrise that was undergoing demolition.

CAUGHT OR TRAPPED

Seven firefighters were killed when they were caught or trapped in 2007. This classification covers firefighters trapped in wildland and structural fires who were unable to escape due to rapid fire progression and the byproducts of smoke, heat, toxic gases, and flame. This classification also includes firefighters who drowned, and those who were trapped and crushed.

- Two Contra Costa County, California, firefighters were killed when fire progressed rapidly in a residential structure.

- Two Boston firefighters were killed when fire progressed rapidly in a restaurant fire.

- Firefighters in Fulton County, Georgia, and Prince William County, Virginia, were killed in residential structure fires where conditions changed rapidly.

- A Baltimore firefighter was severely burned in a training fire in an acquired residential structure.

COLLAPSE

Seven firefighters died in 2007 as the result of structural collapses. This was the fifth most common cause of fatal injury for firefighters in 2007.

- Two Texas firefighters were killed when a structural failure occurred while they were fighting a residential structure fire.

- Firefighters in Indiana and Tennessee were killed when the floors of the residential structures in which they were operating failed, and the firefighters were dropped into the basement of the residence.

- Two firefighters in Pennsylvania (separate incidents) and a firefighter in Nebraska were killed when they were crushed in structural collapses while they were fighting structure fires.

STRUCK BY OBJECT

Being struck by an object was the sixth leading cause of fatal firefighter injuries in 2007. Five firefighters died in 2007 as the result of being struck by an object:

- Two West Virginia firefighters were struck and killed by explosion debris when propane ignited and exploded.

- A Florida chief officer was struck by an unknown object during a structure fire. He received a leg injury that produced a blood clot that resulted in a pulmonary embolism.

- A Florida firefighter was struck by a tree during wildland firefighting training.

- An Illinois firefighter was struck by a passing bus as he worked at a roadside incident.

FALL

Four firefighters died in 2007 as the result of falls. Two firefighters died in separate incidents when they fell from the top of a fire department apparatus; a New York City firefighter fell from the roof of a structure (or from a fire department aerial ladder); an Arkansas firefighter fell to his death from a bridge as he assisted at the scene of a motor vehicle crash.

Enclosed Structure Firefighter Fatalities

A recent analysis of 444 firefighter fatalities that took place while the firefighter was on the scene of a structure fire has revealed the degree of danger associated with open and enclosed structure fires. An open structure is one that possesses numerous openings for firefighter access and egress. An enclosed structure is one with limited openings. Many "big box" retail stores are considered to be "enclosed" structures.

The study also determined the degree of safety provided by the strategy and tactics used during these operations. The analysis considered structural fire firefighter fatalities from January 1, 1990, through December 31, 2006. The study found that 187 or 87 percent of the firefighter fatalities occurred in an enclosed structure, while 36 or 16 percent of the fatalities occurred in an opened structure.

In all cases, an aggressive interior attack was used. Also, 34 (87 percent) of the structure fires resulting in multiple firefighter fatalities occurred in an enclosed structure, while 5 (13 percent) occurred in an open structure.

The analysis concluded that, over a 16-year time span, firefighters using an aggressive interior attack in enclosed structures died far more often, in greater numbers, and with greater multiple line-of-duty deaths than those using the same tactical approach in open structure fires. In response to these significant findings, it is important that departments act to prevent additional firefighter deaths by adopting and implementing more appropriate enclosed structure tactics and Standard Operating Guidelines (SOGs) for use during extremely dangerous enclosed structure fires.

For more information on this work, contact Captain William Mora (Ret.) at capmora@aol.com.

Coming soon to www.usfa.dhs.gov, a USFA SPECIAL REPORT--Killer Buildings: Mitigating the Risks and Defining Tactics for Closed-in Environments.

OTHER

Two firefighters died in 2007 of a cause that is not categorized above. A Kansas fire officer died when he came into contact with an electrical line while fighting a wildland fire. A New York firefighter died of an unknown cause while attending the National Fire Academy (NFA).

NATURE OF FATAL INJURY

Table 10 and Figure 11 show the distribution of the 118 firefighter deaths that occurred in 2007 by the medical nature of the fatal injury or illness.

Table 10: Nature of Fatal Injury (2007)

Nature	Number
Heart Attack	52
Internal Trauma	33
Asphyxiation	18
Burns	7
Crushed	3
CVA	2
Electrocution	1
Other	2
Total	118

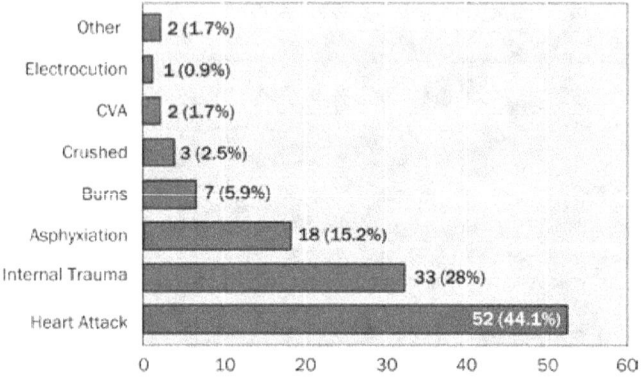

Figure 11. Fatalities by Nature of Fatal Injury (2007).

HEART ATTACK

Heart attacks were the most frequent cause of death for 2007, with 52 firefighter deaths, up slightly from the 50 deaths in 2006, but down from 55 heart attack deaths in 2005, and 61 heart attack deaths in 2004. Figure 12 provides a detailed breakdown of heart attacks by type of duty.

- Fifteen firefighters died of heart attacks that struck after the conclusion of an incident response or onduty period.

  - These firefighters suffered heart attacks within 24 hours of onduty stressful or strenuous activity.

  - The timing of illness onset can range from when the firefighter is walking from the station to his or her vehicle after the conclusion of the incident to the next day.

  - All of the firefighters included in this report under the Hometown Heroes Act criteria are in this group.

- Thirteen firefighters died as the result of heart attacks that occurred while the firefighter was on duty but not assigned to training or response duties.

- Nine heart attacks struck firefighters who were working on the scene of a fire incident. These incidents include firefighters who were engaged in firefighting tasks such as advancing hoselines, includes firefighters on the scene performing support functions such as pump operations.

- Six heart attacks killed firefighters who were engaged in training activities. In 2007, these deaths include two firefighters who were performing physical fitness evaluations and firefighters that became ill while training on firefighting related tasks such as self-contained breathing apparatus (SCBA) drills.

- Five firefighters suffered heart attacks while responding to an emergency in 2007. This includes firefighters who became ill as they responded to the fire station, at the fire station while preparing to respond, and during the response to the incident scene.

- Three firefighters were struck with fatal heart attacks while working at a nonfire emergency in 2007. This included a firefighter working at a hazardous materials emergency.

- One firefighter died in 2007 as the result of a heart attack that occurred as he drove away from a reported structure fire that turned out to be illegal burning.

INTERNAL TRAUMA

In 2007, 33 firefighters died due to internal physical trauma. This grouping includes most firefighters killed in vehicle crashes as well as those who received physical injuries resulting from events such as a building collapse.

- Motor vehicle-related incidents accounted for the deaths of 24 firefighters due to traumatic injuries.

    - Nine firefighters died in crashes involving their personal vehicles: eight while responding to and one while returning from an emergency. While the death of one firefighter resulted from a tree falling on his vehicle during a response, the vast majority of these crashes are at least partially the result of excessive speed. The lack of seatbelt use is also a very common factor in these deaths.

    - Four firefighters died in three incidents involving tankers (tenders). Although tankers make up an estimated 3 percent of the fire service fleet, they are the second most common type of vehicle involved in crashes that result in firefighter fatalities, after personal vehicles.

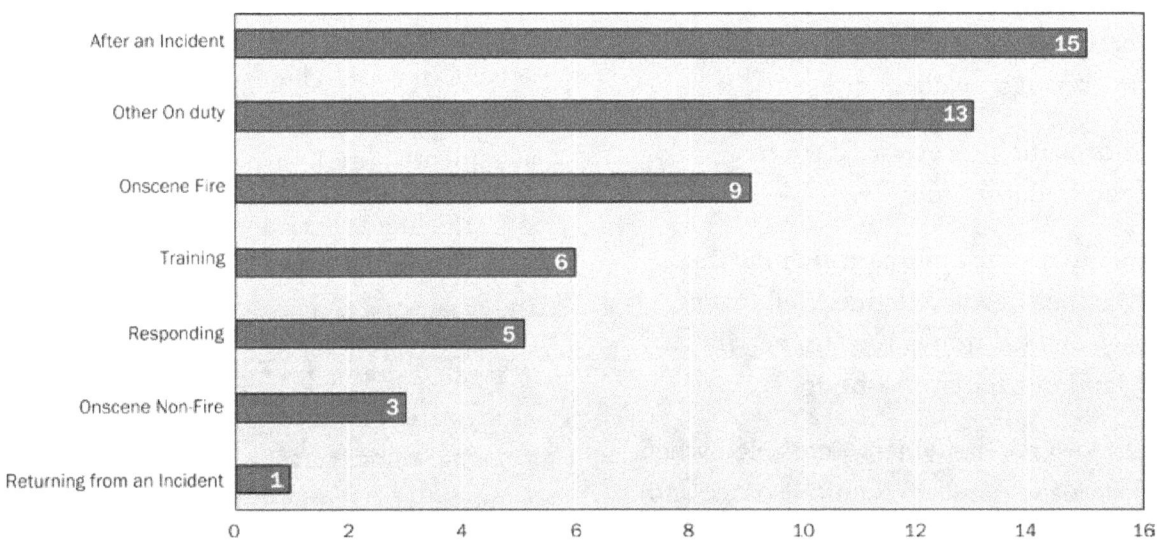

Figure 12. Heart Attacks by Type of Duty (2007).

Table 11: Internal Trauma Firefighter Deaths

Year	Number of Firefighter Deaths
2007	33
2006	24
2005	32
2004	31
2003	41
2002	34
2001	28*
2000	36
1999	25
1998	27

Does not include the firefighter deaths of September 11, 2001, in New York City.

- Three firefighters were killed in fire apparatus crashes involving engine apparatus, and one firefighter died when he was ejected from a fire police van that was involved in a collision.

- Two California chief officers were killed in single-vehicle crashes that involved their government vehicles.

- Three firefighters were killed in crashes that involved their fire department vehicles.

- One firefighter was killed in the crash of a helicopter that contacted trees as it made a supply drop at a wildland fire.

- A California firefighter was killed in a wildland bulldozer rollover.

- Four firefighters died of traumatic injuries received in falls. Two firefighters fell from apparatus, one fell from a roof (or aerial ladder), and one fell from a bridge.

- Two West Virginia firefighters were killed when leaking propane found an ignition source and exploded.

- An Illinois firefighter was killed when he was stuck by a bus as he worked a roadside incident.

- A Pennsylvania firefighter died in the collapse of a burning structure, and a Florida firefighter died as the result of a pulmonary embolism resulting from a fireground injury.

ASPHYXIATION

Asphyxiation was the third leading medical reason for firefighter deaths in 2007, as it was in 2006. Eighteen firefighters died due to asphyxiation in 2007. Sixteen of the deaths were related to structure fires; one involved a drowning; and one involved a vehicle crash.

- Nine Charleston firefighters died of smoke inhalation and thermal injuries while fighting a fire in a retail furniture store.

- Three firefighters died of asphyxiation while fighting residential fires in Georgia, Indiana, and Tennessee.

- Two New York City firefighters died of asphyxiation while fighting a fire in a high-rise structure.

- One of the two Boston firefighters killed in a restaurant fire died of smoke inhalation.

- A Nebraska firefighter died from positional asphyxiation after a structural collapse during a residential fire fight.

- A Utah chief officer drowned after his vehicle veered from the roadway and submerged in the water of a reservoir.

- A South Carolina firefighter died of asphyxiation after he was involved in a personal vehicle crash. The vehicle burst into flames after coming to rest.

Table 12: Firefighter Deaths due to Asphyxiation

Year	Number of Firefighter Deaths
2007	18
2006	12
2005	8
2004	5
2003	6
2002	15
2001	18
2000	13
1999	16
1998	15

BURNS

Seven firefighters died as a result of burns in 2007. Multiple fatality incidents in California and Texas led to the deaths of four firefighters due to burns. One of the two Boston firefighters who died in a restaurant fire died due to burns. A Virginia firefighter killed in a residential fire and a Baltimore firefighter killed during live fire training also died as the result of burns.

CRUSHED

Three firefighters were fatally crushed in 2007: A Florida firefighter was crushed by a fallen tree during training, a Chicago firefighter was crushed under a ladder truck after it was involved in a crash, and a Pennsylvania firefighter was crushed in a structural collapse at a residential fire.

CEREBROVASCULAR ACCIDENT

Two firefighters died in 2007 as a result of strokes (CVAs): One firefighter became ill while responding, and the other firefighter became ill after a protracted physical rescue of an ill person over rough terrain.

ELECTROCUTION

A Kansas firefighter was electrocuted as he fought a wildland fire. He came into contact with a live wire while advancing a hoseline to control a brush fire that was ignited by the fallen line.

OTHER

Two firefighters died in 2007 in situations where the nature of their deaths does not fall into any of the categories described above:

- A New York firefighter died while attending the NFA. The nature of his illness is unknown.

- A Kansas firefighter died of a heart condition that is not considered a heart attack.

Photo by Mark Whitney, U.S. Fire Administration

FIREFIGHTER AGES

Figure 13 shows the percentage distribution of firefighter deaths by age and nature of the fatal injury. Table 13 provides a count of firefighter fatalities by age and the nature of the fatal injury.

In 2007, as in most prior years, younger firefighters were more likely to have died as a result of traumatic injuries such as injuries from an apparatus accident or becoming caught or trapped during firefighting operations. Stress-related deaths are absent below the 31 to 35 years of age category. The youngest firefighter to die of a cardiac-related cause in 2007 was age 34. Statistics show that stress plays an increasing role in firefighter deaths as age increases.

Table 14: Firefighter Ages and Nature of Fatal Injury

Age Range	Nontrauma Total	Trauma Total
under 21	0	5
21 to 25	0	6
26 to 30	0	9
31 to 35	3	6
36 to 40	8	10
41 to 45	8	8
46 to 50	9	8
51 to 60	16	6
61 and over	13	3

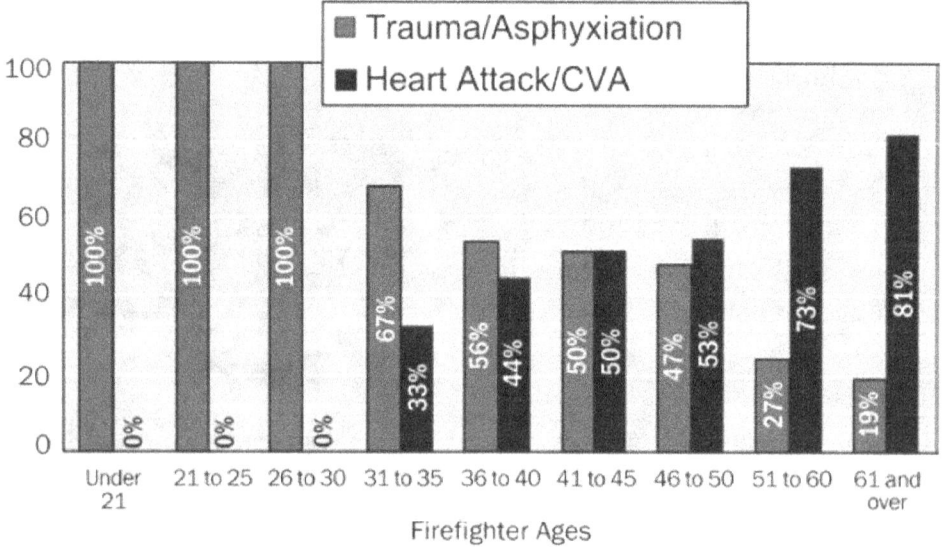

Figure 13. Fatalities by Age and Nature (2007).

The youngest firefighter killed on duty in 2007 was Firefighter Brandon Michael Whimple of North Carolina. He was a passenger in a tanker that was involved in a crash while responding to a structure fire. He was 19 years old.

The oldest firefighter killed on duty in 2007 was Fire Chief Carl Stanley Engdahl of Kansas. He died as the result of a heart attack suffered at a residential fire. He was 78 years old.

FIXED PROPERTY USE FOR STRUCTURAL FIREFIGHTING DEATHS

There were 30 fatalities in 2007 where firefighters became ill or injured while on the scene or engaged in structural firefighting and the fixed property use is known. Table 14 shows the distribution of these deaths by fixed property use. In most years, residential occupancies accounted for the highest number of these fireground fatalities. In 2007, commercial occupancy-related deaths were just as frequent as residential occupancy-related deaths. This is partially accounted for by the deaths of nine Charleston firefighters in a commercial occupancy.

Table 15 shows the number of firefighter deaths in residential occupancies for the past 10 years. Based on NFPA national estimates, fires in residential structures account for approximately 76 percent of all structure fires annually. Civilian fire deaths in residential structures account for approximately 97 percent of all structure fire deaths each year. Historically, the frequency of firefighter deaths in relation to the number of fires is much higher for nonresidential structures.

Table 15: Firefighter Ages and Nature of Fatal Injury

Year	Number of Firefighter Deaths
2007	15
2006	15
2005	18
2004	15
2003	10
2002	21
2001	17
2000	21
1999	23
1998	17

Table 14: Structural Firefighting Deaths by Fixed Property Use in 2007

Fixed Property Use	Number	Percent
Residential	15	50%
Commercial	15	50%

Type of Activity

In 2007, there were a total of 38 firefighter deaths on the fireground. Table 16 and Figure 14 show the types of fireground activities firefighters were engaged in at the time they sustained their fatal injuries or illnesses. This total includes all firefighting duties, such as wildland firefighting and structural firefighting.

Table 16: Type of Activity (2007)

Nature	Number
Fire Attack	19
Search and Rescue	7
Standby and Scene Safety	5
Ventilation	3
Incident Command	2
Water Supply	2
Total	38

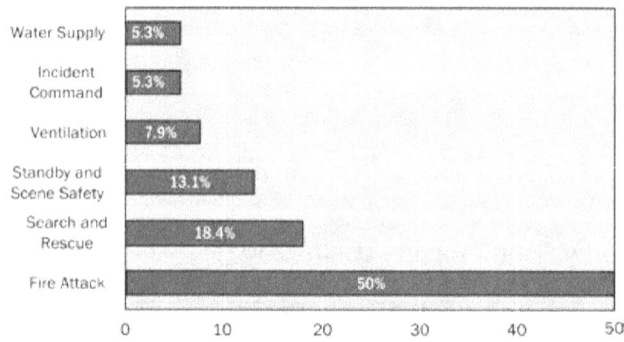

Figure 14. Fatalities by Type of Activity (2007).
Note: Onscene fire only deaths account for 38 of 118 fatalities.

FIRE ATTACK

In 2007, 19 firefighters were killed as they engaged in direct fire attack, such as advancing or operating a hoseline at a fire scene (see Table 17). These include

- Eight of the nine firefighters killed in a retail furniture store fire.
- Two New York City firefighters killed in a highrise fire near Ground Zero.
- Two firefighters who fell into fire-involved basements in separate residential fires.
- Three firefighters (in separate incidents) who were killed in the collapse of structures during fires.
- Two firefighters who suffered heart attacks at structure fires.

Table 17: Firefighter Deaths While Engaged in Fire Attack

Year	Number of Firefighter Deaths
2007	19
2006	24
2005	11
2004	16
2003	11
2002	13
2001	13
2000	13
1999	16
1998	18

- A Boston firefighter who was overcome by fire progress in a restaurant fire.
- A Kansas firefighter who was electrocuted while fighting a wildland fire.

SEARCH AND RESCUE

Seven firefighters were killed in 2007 as they engaged in search and rescue activities:

- Two Contra Costa County, California, firefighters were overcome by fire progress as they searched a residence.

- One of the Boston firefighters killed in a restaurant fire was engaged in searching the structure.

- One of the Charleston firefighters killed in a retail furniture store fire was thought to have been searching for a trapped employee.

- Firefighters in Fulton County, Georgia, and Prince William County, Virginia, were searching residential structures when they were overcome by fire conditions.

- A New York firefighter suffered a heart attack as he searched a fire-involved apartment.

STANDBY AND SCENE SAFETY

Five firefighters died in 2007 while performing standby or scene safety duties at the fire scene:

- Two firefighters suffered heart attacks while on the scene of fire incidents performing support and scene safety duties.

- A helicopter pilot was killed in a crash while delivering supplies.

- A wildland heavy equipment operator lost his life when his bulldozer overturned.

- A Florida chief officer was injured and later died as a complication of his injury after performing support duties on a fire scene.

VENTILATION

Three firefighters died in 2007 while performing ventilation duties on the fire scene: Two Texas firefighters were overcome by fire progress, and a New York City firefighter died in a fall while assigned to ventilation duties.

INCIDENT COMMAND

Two fire chiefs died in 2007 while serving as the Incident Commander for fires in Kansas and Pennsylvania.

PUMP OPERATIONS—WATER SUPPLY

Two firefighters died in 2007 while engaged in water supply duties. Both were operating fire pumpers at structure fires. One death was in Michigan and the other was in Wisconsin.

Photo by Mark Whitney, U.S. Fire Administration

TIME OF INJURY

The distribution of all 2007 firefighter deaths according to the time of day when the fatal injury occurred is illustrated in Figure 15. The time of fatal injury for four firefighters was either unknown or unreported.

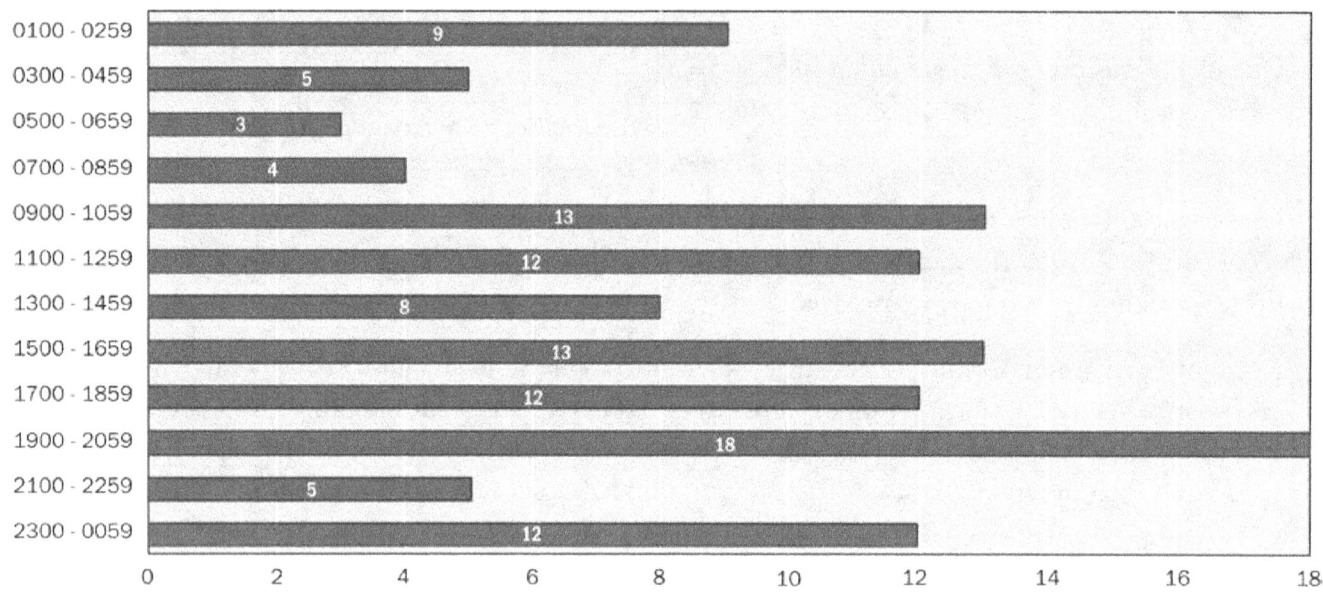

Figure 15. Fatalities by Time of Fatal Injury (2007).

Month of the Year

Figure 16 illustrates the 2007 firefighter fatalities by month of the year.

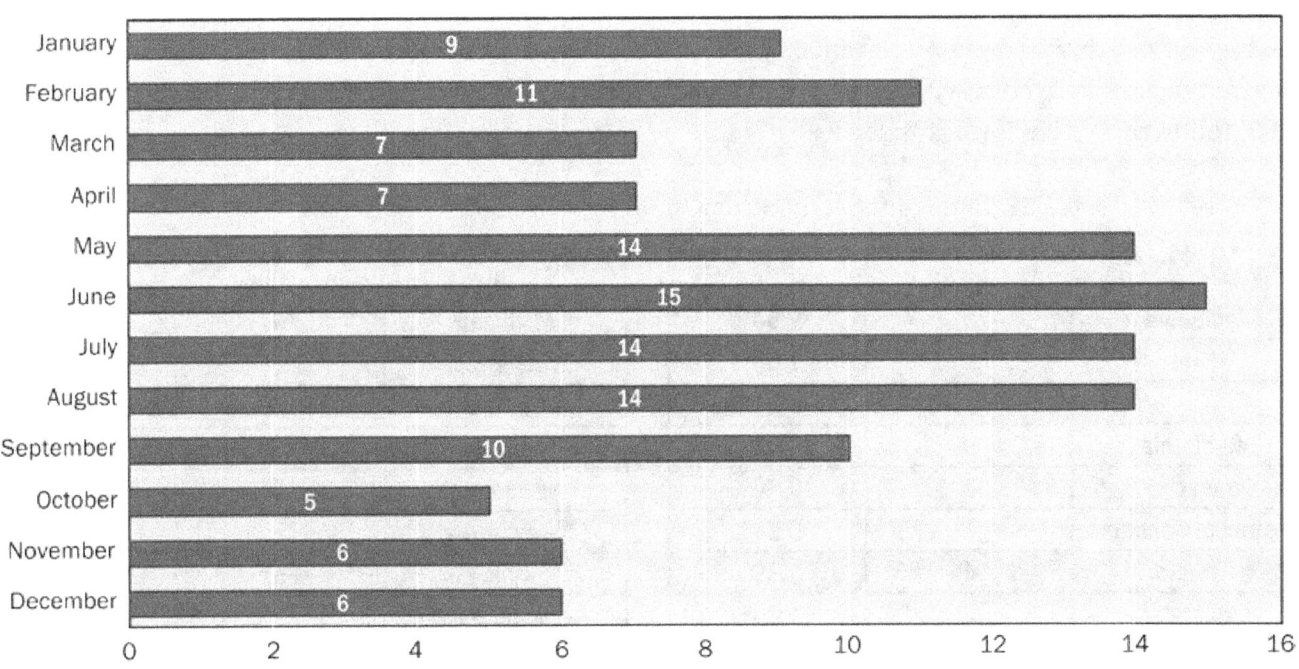

Figure 16. Deaths by Month of Year (2007).

STATE AND REGION

The distribution of firefighter deaths in 2007 by State is shown in Table 18. Firefighters based in 33 States and the District of Columbia died in 2007.

The highest number of firefighter deaths, based on the location of the fire service organization in 2007, occurred in South Carolina with 11 deaths. Pennsylvania had the next highest total of firefighter fatalities at 10, followed by New York at 9 deaths.

Table 18: Firefighter Fatalities by State by Location of Fire Service* (2007)

State	Fatalities	Percentage	State	Fatalities	Percentage
Alabama	3	2.54%	Michigan	5	4.23%
Arkansas	2	1.69%	Mississippi	1	0.84%
Arizona	1	0.84%	North Carolina	8	6.77%
California	9	7.62%	Nebraska	1	0.84%
Connecticut	1	0.84%	New Jersey	5	4.23%
District of Columbia	1	0.84%	New York	9	7.62%
Florida	4	3.38%	Ohio	5	4.23%
Georgia	2	1.69%	Oklahoma	1	0.84%
Idaho	2	1.69%	Pennsylvania	10	8.47%
Illinois	5	4.23%	South Carolina	11	9.32%
Indiana	3	2.54%	Tennessee	3	2.54%
Kansas	5	4.23%	Texas	3	2.54%
Kentucky	3	2.54%	Utah	1	0.84%
Louisiana	1	0.84%	Virginia	1	0.84%
Massachusetts	4	3.38%	Washington	1	0.84%
Maryland	1	0.84%	Wisconsin	2	1.69%
Maine	1	0.84%	West Virginia	3	2.54%

* This list attributes the deaths according to the State in which the fire department or unit is based, as opposed to the State in which the death occurred. They are listed by those States for statistical purposes and for the National Fallen Firefighters Memorial at the NETC.

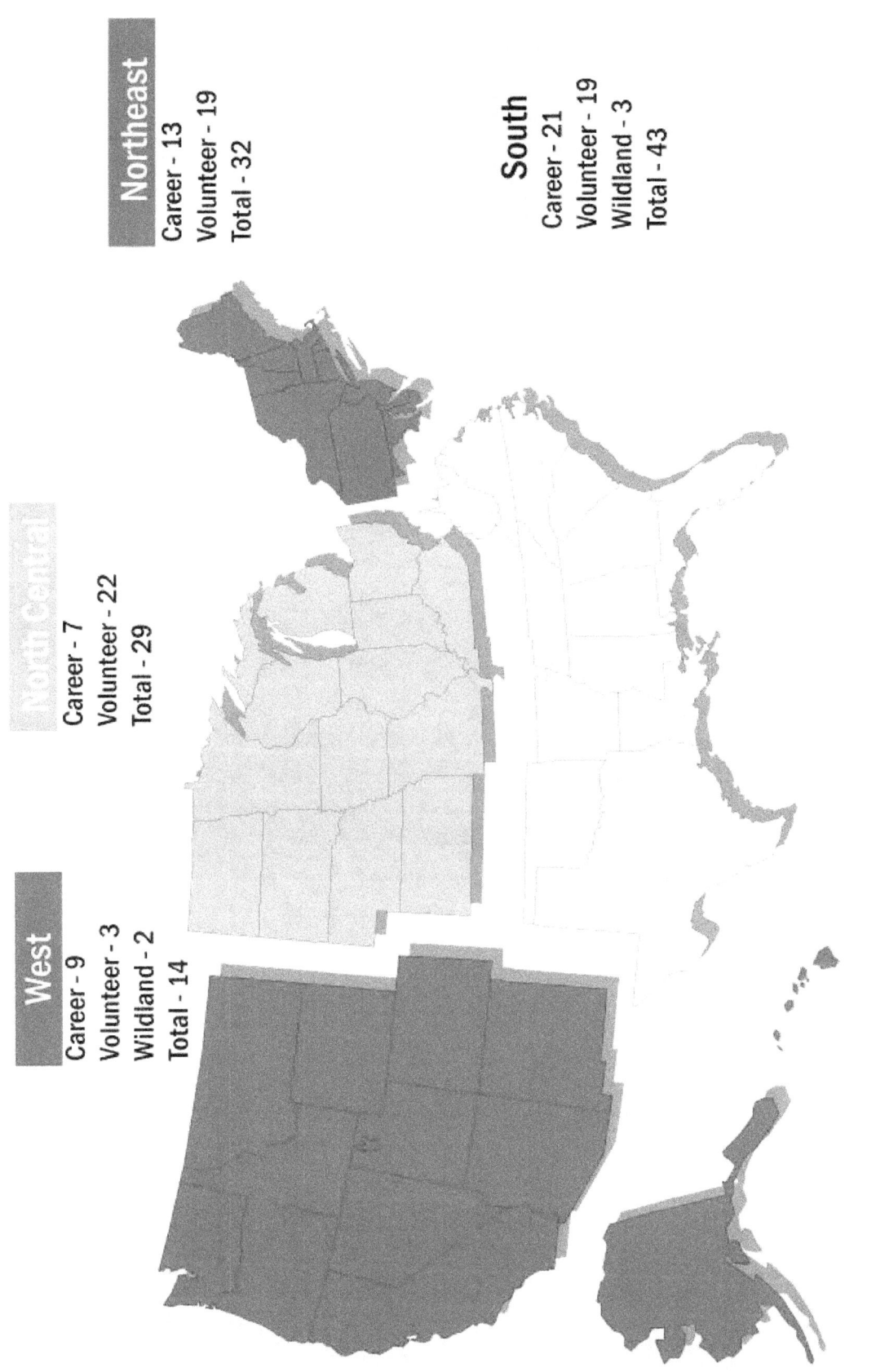

Northeast
Career - 13
Volunteer - 19
Total - 32

South
Career - 21
Volunteer - 19
Wildland - 3
Total - 43

North Central
Career - 7
Volunteer - 22
Total - 29

West
Career - 9
Volunteer - 3
Wildland - 2
Total - 14

Figure 17. Firefighter Fatalities by Region (2007).

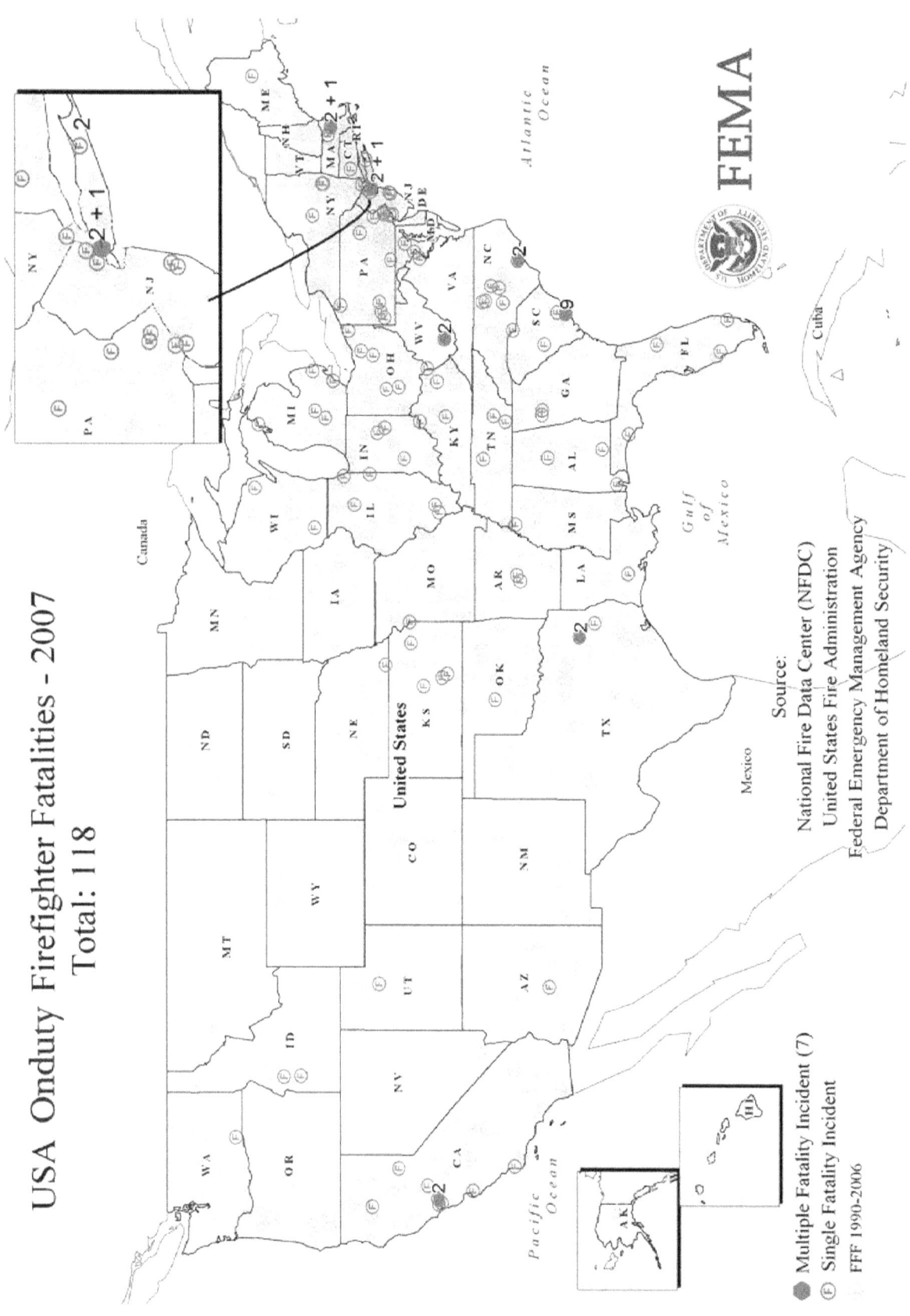

Figure 18. On Duty Firefighter Fatalities 2007 (1990-2006 also shown)

Analysis of Urban/Rural/ Suburban Patterns in Firefighter Fatalities

The United States Census Bureau defines "urban" as a place having a population of at least 2,500 or lying within a designated urban area. "Rural" is defined as any community that is not urban. "Suburban" is not a census term but may be taken to refer to any place, urban or rural, that lies within a metropolitan area defined by the Census Bureau, but not within one of the central cities of that metropolitan area.

Fire department areas of responsibility do not always conform to the boundaries used by the Census Bureau. For example, fire departments organized by counties or special fire protection districts may have both urban and rural coverage areas. In such cases, where it may not be possible to characterize the entire coverage area of the fire department as rural or urban, firefighter deaths were listed as urban or rural based on the particular community or location in which the fatality occurred.

The following patterns were found for 2007 firefighter fatalities. These statistics are based on answers from the fire departments and, when no data from the departments were available, the data are based upon population and area served as reported by the fire departments.

Table 19: Firefighter Deaths by Coverage Area Type

	Urban/Suburban	Rural	Federal or State Parks/Wildland	Total
Firefighter Deaths	64	46	8	118

In Conclusion

In 2007, the families of fallen firefighters, the fire departments that sustained a loss, and the fire service as a whole in the United States suffered another year of unacceptable death and injury.

The 118 deaths in 2007 marked a nearly 11 percent increase in deaths from the 106 suffered in 2006. The lessons that we learn from these tragic incidents serve the fire service best if they are used as the basis of preventing future deaths.

The USFA, NIOSH, the NFFF, and a host of other national fire service organizations are making major efforts to reduce injuries and deaths. None of these efforts will be successful until we change the basic culture of firefighting to one that considers risk more actively. We must not fall into the old adage that injuries and deaths "are part of the business." This is an unacceptable philosophy.

There are many facets to the firefighter fatality problem. Some fatalities could be avoided by actions as simple as slowing down the speed of a response, or wearing a seatbelt. Other problems, such as building construction and firefighter health issues, require a more comprehensive and longer-term solution.

No mission is more important to the fire service than the elimination of needless firefighter deaths. This document seeks to support this effort in some small way.

In memory of all firefighters
who answered their last call in 2007

To their families and friends

To their service and sacrifice

SPECIAL TOPIC—SEATBELTS

In 2007, 27 firefighter fatalities resulted from vehicle-related incidents. In 19 of the non- aircraft incidents where seatbelt status was known, 11 firefighters were not wearing seatbelts at the time of the event. This 2007 statistic continues a disappointing trend from previous years. Fatalities in 2007 where seatbelts were not used were distributed across the volunteer and career fire-EMS service involving multiple vehicle types, under a range of circumstances.

Some examples of 2007 firefighter fatalities where seatbelts were not used are summarized below:

- **March 24, 2007**—Rhodestown (NC) Volunteer Fire Department Firefighters Billy Harold Williams and Brandon Michael Whimple were the driver and passenger, respectively, of a 1,200-gallon water tanker (tender) responding to a reported structure fire.

As the apparatus left a slight left-hand curve, Firefighter Williams braked and steered to the left. The apparatus crossed the center line of the roadway. Firefighter Williams overcorrected/oversteered and the apparatus began to skid sideways. The apparatus left the right side of the roadway, then skidded back onto the roadway and overturned.

Firefighter Whimple was trapped under the vehicle and had to be extricated with a heavy tow truck. Firefighter Williams was ejected in the course of the crash. Neither firefighter was wearing a seatbelt at the time of the crash.

- **May 19, 2007**—Captain John Francis Keane was the Company Officer for Waterbury (CT) Fire Department Engine 8. At 1034 hours on May 19, 2007, Engine 8 and several other fire department units were dispatched to a possible kitchen fire in an apartment. Engine 8 was operating an older, reserve piece of fire apparatus since their regular apparatus was in the shop. In addition, Engine 8 was responding from a location other than their fire station; the crew was attending a blood drive.

Engine 8 and Truck 1 entered an intersection at the same time from different directions. Truck 1 struck the left front of Engine 8. The impact caused the engine to spin clockwise approximately 180 degrees. Captain Keane and the driver of Engine 8 were ejected from the apparatus. The officer on Truck 1 was trapped in the vehicle and had to be extricated. All eight firefighters from Engine 8 and Truck 1 were injured.

Captain Keane was transported to the hospital where he remained until his death on May 22, 2007. The cause of death was listed as blunt head trauma.

A law enforcement investigation of the crash stated that Truck 1 had the green light, that a number of firefighters, including Captain Keane, were not wearing seatbelts at the time of the crash, and that some lines of sight at the crash scene were obstructed by trees.

- **July 2, 2007**—Coal City Community (IN) Volunteer Fire Department Firefighter Dennise Marie Leslie was responding to the fire station in her personal vehicle after her fire department was dispatched to a wildland fire. The vehicle was a 2000 Ford Ranger pickup truck.

As Firefighter Leslie crested a hill, she saw another vehicle waiting to make a left-hand turn in front of her. She moved her vehicle from the center of the road into her lane but oversteered. The right wheels of her vehicle left the paved roadway. Firefighter Leslie steered to the left in an attempt to bring her vehicle back onto the road. The pickup began to spin, travelled across the roadway, and collided with an earthen embankment. This caused the vehicle to go airborne and begin to roll over. The cab of the truck collided with a large tree, causing extensive damage to the passenger compartment.

When EMS responders arrived on the scene, Firefighter Leslie was determined to have died in the crash. The cause of death was listed as head trauma. The pickup's airbag deployed, but Firefighter Leslie was not wearing a seatbelt.

In contrast to these tragic events, there are recent success stories where firefighters' lives were saved by properly wearing seatbelts. As an example,

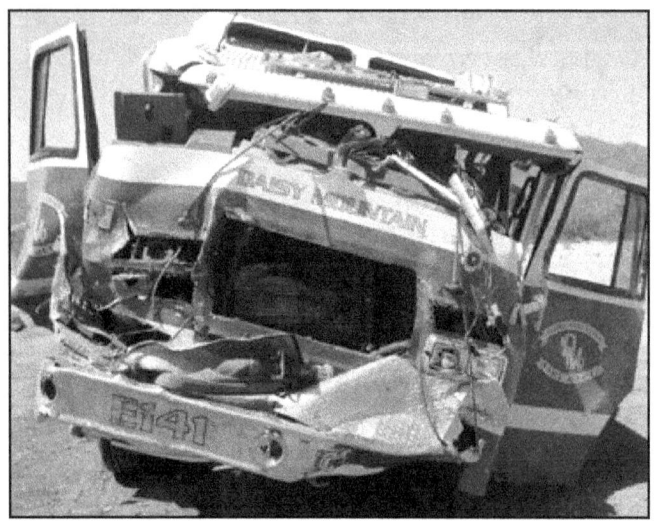

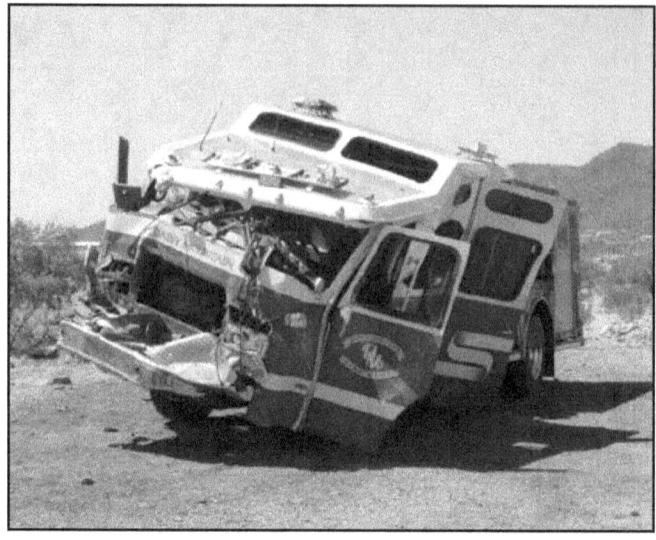

on April 15, 2006, Engine 141 from the Daisy Mountain (AZ) Fire Department was struck by another vehicle on a highway 35 miles north of Phoenix. Despite significant damage to the apparatus (displayed in the photos above and on the left, courtesy of the Daisy Mountain Fire Department), all the firefighters survived with minor injuries; the driver of the other vehicle was killed, despite the firefighters' efforts to provide medical care.

The real tragedy of firefighter fatalities resulting from a lack of seatbelt use is that they are entirely preventable. Fire departments and individual firefighters must recognize this fact and adopt stringent rules, requirements, and cultural norms for wearing seatbelts at all times while riding fire apparatus, operating staff vehicles, or driving privately-owned vehicles (POVs).

First and foremost, all fire apparatus should be equipped with seatbelts, without exception. The 2003 edition of the NFPA 1901, *Standard for Automotive Fire Apparatus*, also requires seatbelts on new apparatus to be "bright red in color" to assist crew members and supervisors with knowing the seatbelt is deployed. (NFPA 1901, 2003, section 14.1.3.1) In addition to "bright red or bright orange" seatbelts, the proposed 2009 edition of NFPA 1901 further requires the installation of audible/visual systems to indicate the status of seatbelts and seat occupancy in fire apparatus. (NFPA 1901 A2008 ROP draft, 2008, sections: 14.1.3.10; 14.1.3.10.1; 14.1.3.10.2; 14.1.3.7.3)

Fire apparatus manufacturers and other stakeholders are exploring other engineering solutions to protect firefighters riding in fire apparatus, such as "roller coaster"-style restraints deployed from overhead.

The 2007 edition of the NFPA 1500, *Standard on Fire Department Occupational Safety and Health Program*, requires that, "All persons riding in fire apparatus shall be seated and belted securely by seatbelts in approved riding positions at any time the vehicle is in motion..." (NFPA 1500, 2007, section 6.3.1) While NFPA 1500 does allow limited exceptions for hose loading (NFPA 1500, 2007, section 6.3.4) and tiller training (NFPA 1500, 2007, section 6.3.5), it requires that fire departments allowing such exceptions do so only after adopting written safety procedures.

There are a number of efforts underway to promote firefighter seatbelt use. One of these is the *Brian Hunton National Fire Service Seat Belt Pledge*, started in 2006 by Dr. Burton Clark of the USFA's NFA, named in honor of a 27-year-old firefighter from Amarillo, Texas, who fell from a fire apparatus on April 23, 2005, and later died from his injuries. A number of fire departments across the country have adopted the seatbelt pledge as an integral part of their firefighter safety efforts. The Dallas (TX) Fire-Rescue Department, for example, recently achieved 100-percent participation in the seatbelt pledge, representing the largest of 200 fire departments across the U.S. to reach this milestone since 2006. (Firehouse.com, 2008)

The Pledge form is reproduced on the following page. More information and additional forms can be accessed from the now *International Seatbelt Pledge* Web site www.trainingdivision.com/seatbeltpledge.asp

Photo by Mark Whitney, U.S. Fire Administration

BRIAN HUNTON: NATIONAL FIRE SERVICE SEAT BELT PLEDGE

Firefighter Christopher Brian Hunton, age 27, was a member of the Amarillo Texas Fire Department for 1 year. On April 23, 2005, he fell out of his fire truck responding to an alarm. He died 2 days later from his injuries. Brian was not wearing his seatbelt.

I pledge to wear my seatbelt whenever I am riding in fire department vehicles. I further pledge to ensure that all my brother and sister firefighters riding with me wear their seatbelts.

I make this pledge willingly; to honor Brian Hunton, my brother firefighter, and because wearing seatbelts it is the right thing to do.

On My Honor, I So Pledge:

Department _____   Address _____

_____

Signature	Print Name	Signature	Print Name
Signature	Print Name	Signature	Print Name
Signature	Print Name	Signature	Print Name
Signature	Print Name	Signature	Print Name
Signature	Print Name	Signature	Print Name
Signature	Print Name	Signature	Print Name
Signature	Print Name	Signature	Print Name

Please fax to the number for the time zone you are in:

(866) 638-3842 Eastern or Pacific; (817) 295-3145 Central; (817) 297-0232 Mountain.

The USFA's *Emergency Vehicle Safety Initiative* (http://www.usfa.dhs.gov/fireservice/research/safety/vehicle.shtm), was developed in conjunction with the U.S. Department of Transportation (DOT) Federal Highway Administration (FHWA), and provides several resources to assist fire departments with ensuring firefighter safety, including seatbelt use.

Working with the International Association of Fire Chiefs (IAFC), USFA developed the *Guide to Model Policies and Procedures for Emergency Vehicle Safety* (http://www.iafc.org/displaycommon.cfm?an=1&subarticlenbr=602), a comprehensive Web-based educational program aimed at reducing the impact of vehicle-related incidents on the fire service and the communities they protect. It provides indepth information for developing policies and procedures required to support the safe and effective operation of emergency vehicles in the fire service, as well as privately-owned vehicles. The Guide includes a model seatbelt policy:

SEATBELT POLICY

Purpose: To establish appropriate and safe behavior regarding the use of safety belts when operating or riding in an emergency vehicle.

Scope: All personnel.

Policy: All persons driving or riding in fire department vehicles shall be seated in approved riding positions with seatbelts or safety restraints fastened at all times when the vehicle is in motion.

The driver shall not begin to move the vehicle until all passengers are seated and properly secured. All passengers shall remain seated and secured as long as the vehicle is in motion. Seatbelts shall not be loosened or released while enroute to dress or don equipment.

Members shall not attempt to mount or dismount from a moving vehicle under any circumstances.

Exception: A fire department member who is providing direct patient care inside an ambulance shall be permitted to release **momentarily** the seatbelt while the vehicle is in motion—**IF IT IS ESSENTIAL TO PROVIDE PATIENT CARE**. When the procedure has been completed, the fire department member shall refasten the seatbelt. Time without the protection of a seatbelt shall be minimized.

Note: NFPA 1500 allows this exception for the ambulance patient compartment; however, effective restraint systems are now available for ambulances. NFPA 1500 also permits exceptions to the seatbelt policy for hose loading and tiller training; however, strict guidelines must be applied to these activities if the exceptions are included in a departmental policy. The fire department should carefully consider whether these exceptions should be included in the departmental policy statement.

As part of this project, the IAFF has also developed a similar innovative Web and computer-based training and educational program—*Improving Apparatus Response and Roadway Operations Safety in the Career Fire Service* (http://www.iaff.org/hs/EVSP/home.html). This comprehensive program, which includes both instructor and participant guides, discusses critical emergency vehicle safety issues such as seatbelt use, intersection safety, roadway operations safety on crowded interstates and local roads, and driver training. As a result of this course, emergency responders will be able to apply basic strategies to safeguard their health and safety while responding to and returning from an incident and while operating on roadways. The IAFF program also contains vehicle safety Standard Operating Procedures (SOPs) addressing emergency vehicle operations, including a section on the use of seatbelts:

All employees are required to use seat belts at all times when operating a Fire Department vehicle. All personnel shall ride only in regular seats provided with seat belts. Riding on tailboards or other exposed positions is not permitted on any vehicle at any time. The company officer and driver of the vehicle shall confirm that all personnel and riders are on-board, properly attired, with seat belts on, before the vehicle is permitted to move. This confirmation shall require a positive response from each rider, as in "ready."

Through this partnership, the NVFC has developed the *Emergency Vehicle Safe Operations for Volunteer and Small Combination Emergency Service Organizations* (http://nvfc.org/page/988/Emergency_Vehicle_Safe_Operations.htm). This innovative Web-based educational program includes an emergency vehicle safety best practices self-assessment, example SOGs, and behavioral motivation techniques to enhance emergency vehicle safety. The NVFC example SOG for seatbelts is reproduced below:

> **Purpose:** To establish positive behavior regarding the use of safety belts when operating or riding in a vehicle.
>
> **Scope:** All personnel.
>
> **Responsibility/Rationale:** To ensure a culture of "buckling up" seat belts before any vehicle movement to prevent death, injury, or property damage.

Key Points to Consider/Include:

• Remain seated and belted anytime the vehicle is in motion

• Buckle up before vehicle moves

• Use belts whether in personal vehicle or department vehicle

• REMEMBER – It's the law.

This page depicts examples of the potential content for SOPs/SOGs for aspects of emergency services operations. It is by no means inclusive on all needs AND SHOULD NOT SIMPLY BE COPIED AND ADOPTED. Use this example to improve or develop your department wide SOPs/SOGs.

Notwithstanding the efforts of many individuals and organizations to help promote firefighters' use of seatbelts at all times, in the final analysis the decision to wear a seatbelt is a personal choice. Every year, an unacceptable number of firefighter fatalities are related to a lack of seatbelt use; fire departments, fire chiefs, Company Officers, and individual firefighters must work together on all fronts to ensure that all firefighters "buckle-up," all the time.

APPENDIX A

Sidney Alan Hall, Firefighter

Jeremy Christopher Adams, Fire Chief

Daryl W. Mutton, Captain

Kevin Charles Reed, Lieutenant

Kenneth Patrick Fahey, Forestry Technician

Shane Michael Daughetee, Firefighter

Anthony T. Catania, Fire Commissioner and Safety Officer

Frederick Allen "Fred" Burroughs, Captain

Craig Lawrence Dorsey II, Firefighter–EMT

Robert H. Hegney, Sr., Lieutenant

Michael L. "Mike" Fox, Firefighter

Shane Todd King, Firefighter

Jeremy Charles LaBella, Firefighter

John W. Broom-Smith, Jr., Firefighter

Joseph Torkos, Fire Engineer Operator

Racheal Michelle Wilson, Firefighter–Paramedic Apprentice

Steven Eric Vanderpool, Jr., Firefighter

Jeffrey M. Murray, Firefighter

Theodore J. Abriel, Sr., Acting Lieutenant

Paul Tyler Reynolds, Sr., Firefighter–EMT

Michael D. Sowich, Firefighter

Lucien Dale Breaux, Captain

Eddy G. Ivers, Fire Chief

William F. Grant, Firefighter

Brandon Michael Whimple, Firefighter

Billy Harold Williams, Firefighter

Steve Olinik, Jr., Firefighter

Edgar Hamlin Scott, Fire Police Captain

Christopher Michael Jaros, Firefighter

Bryan Zollner, North Region Chief of Staff Operations

Kyle Robert Wilson, Firefighter–Technician I

Billy Aaron Gafford, Jr., Captain–Paramedic

Joe Eddy Ivey, Captain

Vernon Robert "Bob" McKenzie, Firefighter

William D. Church, Sr., Fire Police Captain

Bruce Joseph Zumwalt, Firefighter

Peter Beebe-Lawson, Firefighter

Ottis Earl Stephenson, Jr., Firefighter

Brandon Lee Daley, Firefighter

Jeffrey Jeans, Firefighter–EMT

Edward Levirgil Andrews IV, Fire Captain

Joseph E. Piazzi, Deputy Chief and Fire Police Captain

John Francis Keane, Captain

Dennis Cheshire, Firefighter

Bradley William "Wally" Green, Firefighter

Felix Maurice Roberts, Firefighter II

Henry Lawrence "Hank" Pitts, Jr., Lieutenant

David A. Middleton, Firefighter

Mark Stevens Carter, Engineer–Paramedic

David Allan Rufer, Firefighter

Theodore Michael Benke, Captain

William H. "Billy" Hutchinson, Captain

Louis Mark Mulkey, Captain

Mark Wesley Kelsey, Acting Captain

Bradford Rodney "Brad" Baity, Engineer

Michael Jonathon Alan French, Assistant Engineer

Melvin Edward Champaign, Firefighter

James Allen "Earl" Drayton, Firefighter

Brandon Kenyon Thompson, Firefighter

Daniel F. Pujdak, Firefighter

Timothy Lavern Sanborn, Firefighter

Edward Charles Summers, Firefighter

Samuel W. Downing, Captain

Michael B. Douthitt, Driver–Engineer

Ronald Yale Wiley, Deputy Fire Marshal

Dennise Marie Leslie, Firefighter

Eric Robert Lyons, Firefighter

Michael James Penovich, Fire Chief

James J. McRae III, Firefighter–Technician

Stephen R. Dembski, Firefighter

Matthew Charles Burton, Captain

Scott Peter Desmond, Fire Engineer

Cornelius Myron "Storm" Nolton, Firefighter

Dennis Luster Davis, Pilot

Jon Charles Trainer, Firefighter

James Glenn "Shib" Miller, Firefighter

Nemeth Fitzhugh Sanders, Forest Fire Equipment Operator

Kevin Glenn Williams, Captain

Austin Hague Cheek, Firefighter

Gerald David "Jerry" Donley, Jr., Deputy Fire Chief

Paul Darrin Baker, Lieutenant

Todd Whitney Hage, Firefighter

Anthony Philip Cox, Captain

Glenn Williams Miller, Probationary Firefighter

Michael Paul Stephen Heuer, Fire Chief

Robert C. Beddia, Firefighter

Joseph Graffagnino, Firefighter

Jeffrey Dale Swartz, Firefighter

M.L. Hopper, Forestry Aide 2

Paul J. Cahill, Firefighter

Warren J. Payne, Firefighter

George H. Crotts, Jr., Engineer

Jared W. Zimmerly, Firefighter

Michael Dean Stanfield, Captain

Leonard R. Bailey, Jr., Fire Chief

Jerry Wayne Stucker, Firefighter–EMT

Kelly L. Page, Firefighter

William D. "Billy" McDaniels, Sr., Fire Police Captain

Bryon Wayne Johnson, Lieutenant

John E. "Dawg" Lietzke, Firefighter

Mike D. Reagan, Jr., Firefighter

Adam Edward Cole, Firefighter

Matthew Richard Will, Heavy Equipment Operator

Robert Wayne Phillips, Firefighter

Ralph M. Cross, Assistant Fire Chief

Scott A. Mumm, Lieutenant

Jeremy W. Wach, Firefighter

Carl Stanley Engdahl, Fire Chief

Ronny Allen Bennett, Fire Chief

Michael J. Tluscik, Sr., Senior Firefighter

John Jacob "JJ" Curry, Firefighter

Jon Mark Bingham, Assistant Fire Chief

Donald Wallis, Firefighter

Alphonse Vincent Germano, Jr., First Assistant Chief

Peter G. Neilson, Firefighter–Emergency Medical Responder

Raymond Charles Simonis III, Firefighter

Theresa Maria Lynn, Assistant Captain

Walter C. Fagan, Jr., Firefighter

January 3, 2007–1535 hrs
Sidney Alan Hall, Firefighter
Age 52–Volunteer
Upland Volunteer Fire Department, Indiana

The Upland Volunteer Fire Department was dispatched to a structure fire at 1521 hours. Firefighter Hall responded on an engine.

Upon their arrival, firefighters found a two-story wood-frame structure with heavy smoke showing. There was no fire visible from the exterior.

Firefighter Hall was by himself on the nozzle of an attack line a short distance into the structure. He began to flow water on visible fire. As he rotated to apply water, his feet fell through the floor. Firefighter Hall held himself at floor level with his arms but was unable to get out of the hole.

Firefighters entered the structure in an attempt to free Firefighter Hall from the hole but were unable to do so due to smoke conditions and the lack of self-contained breathing apparatus (SCBA) for some firefighters. A firefighter with an SCBA held on to Firefighter Hall to prevent him from falling into the basement. A large piece of plaster fell and struck this firefighter on the head. The firefighter lost his hold on Firefighter Hall and Firefighter Hall fell into the basement.

Firefighters lowered a ground ladder into the hole and entered the basement. After some difficulty, Firefighter Hall was removed from the basement and brought to a waiting ambulance. Approximately 20 minutes had passed between the time Firefighter Hall fell into the hole and when he was removed from the structure.

Firefighter Hall was transported by ambulance and medical helicopter to a regional hospital. Despite treatment at the hospital, Firefighter Hall died on January 5, 2007. The cause of death was listed as positional asphyxiation.

January 9, 2007–1230 hrs
Jeremy Christopher Adams, Fire Chief
Age 40–Career
Springfield Fire Department, Florida

On January 9, 2007, Chief Adams and the members of his fire department responded to a mutual-aid structure fire in a neighboring jurisdiction. When he returned to his fire station, Chief Adams complained of pain in his leg as the result of an injury he received while fighting the fire. He did not seek medical attention for his injury. Over the next several days, Chief Adams continued to complain about his leg and also began to complain of shortness of breath. He dismissed the shortness of breath as the onset of a chest cold.

continued on next page

On January 16, 2007, at approximately 0930 hours, Chief Adams collapsed at City Hall and was taken to the hospital. He was treated in the ambulance and in the hospital, but died. No autopsy or toxicological analysis were performed. A death certificate worksheet prepared by the medical examiner attributed the death to a pulmonary embolism that resulted from a blood clot associated with the previous leg injury.

January 19, 2007–1906 hrs
Daryl W. Mutton, Captain
Age 47–Volunteer
North Pulaski Fire Protection District #5, Arkansas

Captain Mutton and the members of his fire department responded to a fire alarm in a residence. The alarm was unintentional, and firefighters cleared the scene at approximately 1825 hours. At approximately 1900 hours, Captain Mutton's family members called the fire department looking for Captain Mutton. He had not been seen since responding to the alarm.

Captain Mutton was found in his personal vehicle along the side of the road in cardiac arrest. He was taken to the hospital, but was not revived. His death was caused by a heart attack.

January 20, 2007–0830 hrs
Kevin Charles Reed, Lieutenant
Age 47–Career
Oakland Fire Department, California

Lieutenant Reed completed a work shift. During the shift, Lieutenant Reed responded to six incidents, including a residential structural fire. After getting off work, Lieutenant Reed went to a local gym to participate in physical fitness activities.

While riding an exercise bike, Lieutenant Reed collapsed. Other members of the gym began cardiopulmonary resuscitation (CPR) and the gym staff applied an automatic external defibrillator (AED). Firefighters arrived and took over the care of Lieutenant Reed. He was transported to the hospital, but was not revived.

Lieutenant Reed's cause of death was cardiac-related.

January 25, 2007–2000 hrs
Kenneth Patrick Fahey, Forestry Technician
Age 35–Wildland Full-Time
United States Department of Agriculture, Forest Service, South Carolina

Forestry Technician Fahey participated in a controlled burn. He was relieved by other firefighters and released to return to his base station.

As he drove his unit back to base, the vehicle operated by Forestry Technician Fahey crossed the center line of the road for unknown reasons and struck an oncoming semi-truck. Both vehicles were estimated to have been travelling at 55 miles per hour. After impact, both vehicles burst into flames. Forestry Technician Fahey and the driver of the truck were killed.

Forestry Technician Fahey was driving a 3/4-ton pickup with dual rear wheels and a slip-in brush pump and tank assembly. The water capacity of the water tank was 200 gallons. Forestry Technician Fahey was wearing his seatbelt at the time of the crash.

The autopsy listed the cause of death as head/neck blunt force trauma.

January 26, 2007–0059 hrs
Shane Michael Daughetee, Firefighter
Age 24–Volunteer
Highway 58 Volunteer Fire Department, Inc., Tennessee

Firefighter Daughetee and the members of his fire department were dispatched to a report of a structure fire related to a wood stove in the basement of a residential structure. The first firefighter on the scene was the Incident Commander (IC), who observed fire coming from a garage door at the corner of the basement.

Firefighter Daughetee arrived on the first engine. A second engine established a water supply from a fire hydrant. Firefighter Daughetee and other firefighters entered the structure with a handline and encountered zero visibility conditions. A thermal imaging camera (TIC) and ventilation were requested. A positive pressure ventilation (PPV) fan was set up in the doorway through which firefighters had entered. Firefighter Daughetee exited the structure to retrieve the TIC. When he returned, heat conditions had worsened and the floor was described as "spongy." Firefighters withdrew from the structure.

Firefighter Daughetee requested the nozzle and proceeded back into the structure. The floor below Firefighter Daughetee collapsed and Firefighter Daughetee fell into the basement.

A firefighter who had been behind Firefighter Daughetee reached down into the hole, but Firefighter Daughetee was unable to reach his hand. Firefighters placed an attic ladder into the hole and called to Firefighter Daughetee to climb out. By this time, Firefighter Daughetee had become disoriented and was unable to assist with his own rescue. Firefighters entered the basement and made contact

with Firefighter Daughetee but were unable to retrieve him due to limited air supplies and fire conditions.

The structure became unstable and firefighters were withdrawn from the building. Firefighter Daughetee's remains were recovered by firefighters after the fire was controlled. The cause of death was listed as inhalation of smoke and toxic products of combustion.

The floor assembly that collapsed was later determined to be constructed with engineered lumber. A number of firefighter fatalities have occurred in collapses involving these materials.

For additional information regarding this incident, please refer to NIOSH Fire Fighter Fatality Investigation and Prevention Program report F2007-07 (http://www.cdc.gov/niosh/fire/reports/face200707.html).

January 28, 2007–1508 hrs
Anthony T. Catania, Fire Commissioner and Safety Officer
Age 76–Volunteer
Middle Island Fire Department, New York

Fire Commissioner Catania and the members of his fire department were dispatched to an emergency medical incident. As he was leaving his home for the response, Fire Commissioner Catania began to feel ill and called for other firefighters to respond to his residence.

Firefighters found Fire Commissioner Catania in the front seat of his car complaining of difficulty breathing and numbness. He was exhibiting signs of a cerebrovascular accident (CVA) (stroke). He was transported by fire department ambulance to the hospital, where he underwent surgery, and was transferred to an intensive care unit. Fire Commissioner Catania did not recover from his illness. He died on February 2, 2007.

Fire Commissioner Catania responded to a total of four incidents in the 24-hour period prior to the onset of his illness.

January 30, 2007–1053 hrs
Frederick Allen "Fred" Burroughs, Captain
Age 51–Volunteer
Ghent Area Volunteer Fire Department, West Virginia

Craig Lawrence Dorsey II, Firefighter–EMT
Age 24–Volunteer
Ghent Area Volunteer Fire Department, West Virginia

The propane gas service for a local convenience store was being transferred from one provider to another. Service technicians from the new provider were at the store to install the new propane tank, transfer propane from the old tank to the new tank, and establish service for the store.

At 1042 hours, firefighters were dispatched to the report of a gas leak at a local convenience store. Firefighter Dorsey was on duty and responded as a part of an ambulance crew. Captain Burroughs was working his regular job as a county building inspector and responded in his county vehicle.

Firefighters arrived on the scene to find an uncontrollable propane leak. Firefighters reported seeing a billowing cloud of vapor or mist that was striking the eaves of the building and traveling along the ground. An evacuation was begun.

An explosion occurred at approximately 1053 hours. Captain Burroughs, Firefighter Dorsey, and two gas company employees were killed in the blast. Five other people received serious injuries.

The cause of death for both Captain Burroughs and Firefighter Dorsey was listed as trauma.

An investigation by the United States Chemical Safety Board is ongoing. Preliminary work suggests a valve malfunction contributed to the leak and subsequent explosion. Additional information about this investigation can be found at http://www.csb.gov

February 1, 2007–0130 hrs
Robert H. Hegney, Sr., Lieutenant
Age 53–Volunteer
Se-Wy-Co Volunteer Fire Company, Pennsylvania

Lieutenant Hegney responded to a mutual-aid residential structure fire as the driver of his fire department's ladder truck on February 1, 2007. The ladder company was not used at the scene, but was assigned to Staging. The unit was released from the scene at 1708 hours. That evening, at the fire company's weekly drill, Lieutenant Hegney told other firefighters that he was feeling ill.

Lieutenant Hegney performed a number of maintenance and administrative activities at the fire station on February 2. He again complained of not feeling well.

At approximately midnight, Lieutenant Hegney went to the hospital complaining of heart attack symptoms. He was evaluated and brought to a heart catheterization lab for treatment. Lieutenant Hegney died during the procedure. The cause of death was listed as a heart attack.

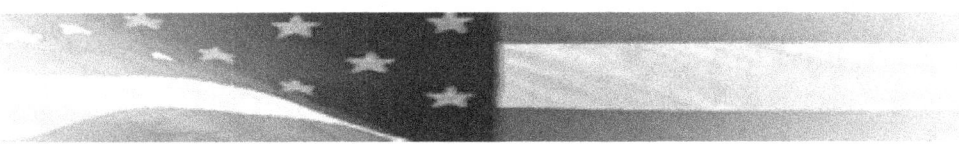

February 2, 2007–0025 hrs
Michael L. "Mike" Fox, Firefighter
Age 37–Volunteer
Vergennes Fire Department, Illinois

Firefighter Fox was driving a fire department ambulance responding to a medical emergency. Firefighter Fox began to experience difficulty breathing but was able to pull the ambulance safely to the side of the road before becoming unconscious.

The firefighter who had been riding as the passenger in the ambulance provided CPR, and Firefighter Fox was transported to the hospital. He was pronounced dead due to a heart attack at 0109 hours.

February 3, 2007–1400 hrs
Shane Todd King, Firefighter
Age 29–Career
Danville Fire Department, Kentucky

Firefighter King was returning from fire department-mandated emergency medical training in a neighboring county. He was driving his personal vehicle, a 1998 Ford Ranger pickup.

As Firefighter King drove north on a local highway, he crossed the center line of the highway for reasons unknown and struck an oncoming tractor-trailer rig.

Firefighter King was wearing his seatbelt at the time of the crash, and his vehicle's airbag deployed. He was trapped in the vehicle and had to be extricated. Firefighter King died at the scene of multiple blunt trauma.

February 4, 2007–0931 hrs
Jeremy Charles LaBella, Firefighter
Age 27–Career
Washington Fire Department, Pennsylvania

Firefighters from the Washington Fire Department and several mutual-aid fire departments responded to a fire in an old motorcycle shop. Firefighter LaBella was a part of the first crew to arrive on the scene. Heavy black smoke was showing from the building.

Firefighter LaBella and another firefighter donned their structural firefighting personal protective equipment (PPE), including SCBA, and advanced an attack line into the garage area of the structure. Additional handlines were deployed, but the fire was not controlled. The IC made preparations for a switch to a defensive strategy.

continued on next page

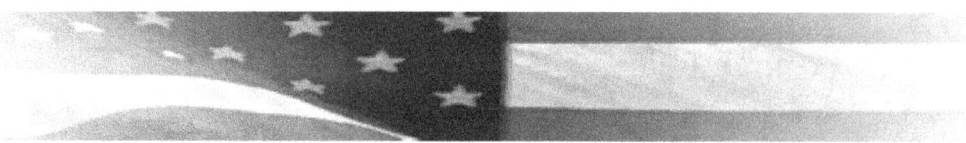

Firefighter LaBella and his partner had exchanged SCBA cylinders and re-entered the structure. A structural collapse of an awning trapped Firefighter LaBella and his partner in the rubble. An onscene Rapid Intervention Crew (RIC) was deployed immediately, and removed both firefighters within 15 minutes of the collapse.

Firefighter LaBella was transported to the hospital by ambulance but succumbed to his injuries. The cause of death was listed as asphyxiation due to entrapment.

For additional information regarding this incident, please refer to NIOSH Fire Fighter Fatality Investigation and Prevention Program report F2007-08 (http://www.cdc.gov/niosh/fire/reports/face200708.html).

February 4, 2007–1300 hrs
John W. Broom-Smith, Jr., Firefighter
Age 44–Volunteer
Sea Side Heights Volunteer Fire Department, New Jersey

Firefighter Broom-Smith and the members of his fire department responded to an automatic fire alarm late on the morning of February 4. Nothing was found, and the fire alarm was reset. Firefighters returned to the fire station at 1127 hours.

Upon returning to the fire station, Firefighter Broom-Smith and approximately five other firefighters removed the fire department's Jet Ski® from a trailer and then unpacked and placed a new Jet Ski® on the trailer. The old Jet Ski® weighed approximately 600 pounds, and the new one, a larger model, weighed approximately 700 pounds. Firefighters strained to carry the heavy watercraft and had to stretch to get the units on and off of the trailer.

At approximately 2300 hours on February 4, 2007, Firefighter Broom-Smith suffered a fatal heart attack.

February 7, 2007–1830 hrs
Joseph Torkos, Fire Engineer Operator
Age 47–Career
Detroit Fire Department, Michigan

Fire Engineer Torkos and his engine company were responding to a report of a structure fire in a vacant residence. Fire Engineer Torkos was driving with the apparatus lights and siren operating. The engine was staffed by an officer, Fire Engineer Torkos, and two firefighters.

continued on next page

As the apparatus responded, it was struck by a Chevrolet Tahoe SUV that approached from the right side of the apparatus at high speed. The force of the collision deflected the path of the apparatus and it ended up off the opposite side of the roadway.

Fire Engineer Torkos and the Company Officer were ejected from the apparatus. Fire Engineer Torkos was ejected through the driver's door or the windshield. He was run over by the apparatus, trapped under the vehicle's rear wheels, and died at the scene. Firefighters were unable to remove Fire Engineer Torkos until a large tow truck arrived at the scene.

Law enforcement reports indicate that the SUV disregarded the flashing traffic control light at the intersection and the warning lights on the apparatus. The SUV was estimated to have been travelling approximately 80 miles per hour at the time of the impact. The engine's speed was estimated at 15 to 30 miles per hour.

The other three firefighters on the engine were injured. The driver of the SUV died in the crash, and three passengers in the SUV were injured. The cause of the structure fire was found to be arson.

February 9, 2007–1200 hrs
Racheal Michelle Wilson, Firefighter–Paramedic Apprentice
Age 29–Career
Baltimore City Fire Department, Maryland

Firefighter Wilson and the members of her fire academy class were attending a live fire training exercise in a vacant rowhouse in Baltimore.

Firefighter Wilson was assigned to a group of apprentices and an instructor designated as Engine 1. Her group advanced a dry attack line into the structure. As they climbed the stairs, the line was charged. Engine 1 encountered and extinguished fire on the second floor but did not check the rest of the second floor for fire prior to proceeding to the third floor. On the third floor, they again encountered and began to extinguish fire.

Fire conditions began to worsen with a marked increase in smoke and heat that appeared to be coming from the second floor. Engine 1 firefighters who were on the stairs began to receive burns from the fire conditions. The instructor for Engine 1 climbed out a window at the top of the stairs and helped one burned firefighter escape to the roof.

Firefighter Wilson appeared at the window in obvious distress and attempted to escape. The windowsill was unusually high (41 inches) and she was unable to escape. Firefighter Wilson momentarily moved away from the window, at which time she advised other firefighters to go down the stairs to escape. When she returned to the window, her SCBA facepiece was off and she was beginning to receive burns. She was able to get her upper body out of the window but she could not make it through. Firefighters on the exterior were unable to pull her through until firefighters were able to gain access on the interior and assist with the effort.

continued on next page

When Firefighter Wilson was pulled to the roof, she was in full cardiac and respiratory arrest. She was immediately removed from the roof and received advanced life support care and transportation to the hospital. She was pronounced dead at 1250 hours.

Firefighter Wilson received total body surface burns of 50 percent. The cause of death was listed as thermal burns and asphyxiation.

February 11, 2007–1639 hrs
Steven Eric Vanderpool, Jr., Firefighter
Age 28–Volunteer
White Oak Volunteer Fire Department, Kentucky

Firefighter Vanderpool was responding in his personal vehicle to a report of a vehicle crash. Firefighter Vanderpool was operating a 1997 Ford Ranger pickup.

Firefighter Vanderpool approached a T-intersection, failed to stop for a stop sign, and proceeded into the intersection. As his vehicle passed through the intersection, it was struck by a 1998 Chevrolet 1500 pickup. The force of the collision spun Firefighter Vanderpool's vehicle around. In the course of the spin, the driver's door opened and Firefighter Vanderpool was ejected. Firefighter Vanderpool was not wearing his seatbelt at the time of the crash.

Firefighter Vanderpool received serious head injuries in the crash. He was taken by ground ambulance and then by helicopter to a regional trauma center, but he never regained consciousness and was pronounced dead on February 24, 2007.

February 13, 2007–2100 hrs
Jeffrey M. Murray, Firefighter
Age 40–Volunteer
Sharon Township Fire Department, Ohio

Firefighter Murray responded to the fire station to operate the radio while other firefighters were attending an emergency medical services (EMS) response. The area was experiencing severe winter weather, with accumulation of approximately 10 inches of heavy snow.

Firefighter Murray stayed at the fire station after the incident was concluded to shovel snow from the sidewalks and other areas that could not be reached by a plow. At approximately 1719 hours, Firefighter Murray returned home, where he also shoveled snow.

Firefighter Murray complained of not feeling well. He went upstairs in his home and collapsed. Firefighters were called to the home at 2114 hours and found him in full arrest. Firefighter Murray was transported to the hospital, but did not survive the heart attack.

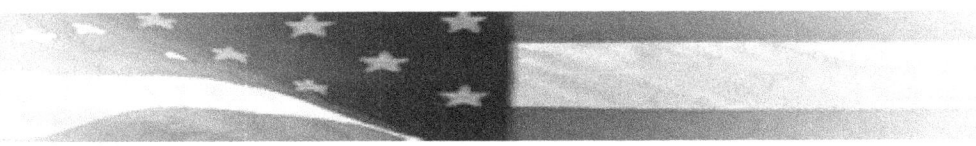

February 19, 2007–1740 hrs
Theodore J. Abriel, Sr., Acting Lieutenant
Age 44–Career
Albany Fire Department, New York

Acting Lieutenant Abriel was the Company Officer on the rescue squad for the shift beginning at 0800 hours on February 19. The crew conducted standard activities around the station and participated in a cancer screening. The rescue squad responded to a structural fire in a residence at 1227 hours. Acting Lieutenant Abriel and his crew participated in the fire fight in a two-story structure.

At 1642 hours, the rescue squad responded to a fire in a residential highrise. Acting Lieutenant Abriel climbed to the sixth floor of the building and participated in firefighting duties for approximately 30 minutes. While working, he suddenly collapsed.

Firefighters moved Acting Lieutenant Abriel to the stairwell and began CPR. Firefighter-paramedics attached a cardiac monitor. Acting Lieutenant Abriel was shocked and provided with other advanced life support services. While his pulse returned briefly, he was not revived.

February 27, 2007–Time Unknown
Paul Tyler Reynolds, Sr., Firefighter–EMT
Age 41–Career
Estero Fire Rescue, Florida

Firefighter Reynolds had worked a 24-hour shift, during which he and his crew responded to three emergency incidents, performed hydrant maintenance and testing, and performed fire station and apparatus maintenance duties. Although the temperatures of the day were moderate, it was very humid.

Firefighter Reynolds completed his shift and went home. At approximately 1300 hours, Firefighter Reynolds collapsed. Firefighters were called to the scene and provided paramedic-level care at the scene and en route to the hospital. Despite aggressive treatment efforts by firefighters and hospital personnel, Firefighter Reynolds did not survive his heart attack.

March 2, 2007–2030 hrs
Michael D. Sowich, Firefighter
Age 50–Volunteer
New Hartford Fire Department, New York

Firefighter Sowich was attending the New York State Weekend at the National Fire Academy (NFA) in Emmitsburg, Maryland. While he had been at New Hartford for only a short time, he had been a long-time member of the Sherrill-Kenwood Fire Department.

Firefighter Sowich was discovered deceased in his dorm room. The cause of his death is unknown.

March 13, 2007–1725 hrs
Eddy G. Ivers, Fire Chief
Age 59–Volunteer
Concord-Greene Township Volunteer Fire Department, Ohio

Chief Ivers and the members of his fire department were dispatched to a possible barn fire in their community. Chief Ivers drove a pumper. When firefighters arrived on the scene, they learned that the fire was a controlled burn. Chief Ivers explained the community's open burning laws to the home owner, and firefighters cleared the scene to return to the station.

A firefighter driving behind the apparatus driven by Chief Ivers saw the vehicle slow and pull to the side of the road. The firefighter assumed that the apparatus had broken down and went to the cab of the apparatus to offer assistance. As the firefighter arrived, Chief Ivers was getting out of the cab of the truck and said that he was ill. The firefighter summoned EMS assistance, and several passers-by offered to help.

Chief Ivers was transported to the hospital by ambulance and later was flown by air ambulance to a regional care facility, where he died as the result of a heart attack.

March 17, 2007–0045 hrs
Lucien Dale Breaux, Captain
Age 57–Career
Crowley Fire Department, Louisiana

Captain Breaux was on duty in his assigned fire station. He was asleep and was awakened by the dispatch of other firefighters to a vehicle extrication.

Captain Breaux was not feeling well and contacted dispatch to request an emergency medical response to his location. Firefighters and the local ambulance service were dispatched to the station and found Captain Breaux having difficulty breathing and complaining of chest pain.

continued on next page

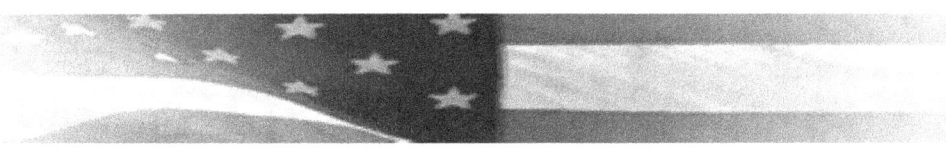

Paramedics evaluated Captain Breaux, and he was transported to the hospital. Captain Breaux went into cardiac arrest twice during his transport and was ultimately pronounced dead in the hospital emergency room at 0149 hours.

March 23, 2007–0741 hrs
William F. Grant, Firefighter
Age 44–Career
Chicago Fire Department, Illinois

Firefighter Grant was a passenger in a ladder truck responding to a reported structure fire. Firefighter Grant was in the rear seat of the apparatus behind the driver. As the apparatus proceeded through a red light, it was struck from the right by a school bus. The point of impact was the right rear area of the ladder truck. After the impact, the apparatus spun in a clockwise direction and landed on the driver's side.

Firefighter Grant was partially ejected from the apparatus and pinned under the apparatus when it came to rest. Firefighter Grant was extricated by other firefighters and transported to the hospital. He was pronounced dead upon arrival. Firefighter Grant died as the result of multiple blunt force injuries.

March 24, 2007–1159 hrs
Brandon Michael Whimple, Firefighter
Age 19–Volunteer
Rhodestown Volunteer Fire Department, Inc., North Carolina

Billy Harold Williams, Firefighter
Age 45–Volunteer
Rhodestown Volunteer Fire Department, Inc., North Carolina

Firefighter Williams and Firefighter Whimple were the driver and passenger, respectively, of a 1,200-gallon water tanker (tender) responding to a reported structure fire.

As the apparatus left a slight left-hand curve, Firefighter Williams braked and steered to the left. The apparatus crossed the center line of the roadway. Firefighter Williams overcorrected/oversteered, and the apparatus began to skid sideways. The apparatus left the right side of the roadway, then skidded back onto the roadway and overturned.

continued on next page

Firefighter Whimple was trapped under the vehicle and had to be extricated with a heavy tow truck. Firefighter Williams was ejected in the course of the crash. Neither firefighter was wearing a seatbelt at the time of the crash.

March 29, 2007–1723 hrs
Steve Olinik, Jr., Firefighter
Age 65–Volunteer
Rome Volunteer Fire Department, Ohio

Firefighter Olinik and the members of his fire department had responded earlier in the day to emergency incidents. At approximately 1723 hours, his department was dispatched on a mutual-aid wildland incident.

When Firefighter Olinik arrived at the fire station, the first apparatus already had departed for the incident scene. Other firefighters stood by at the fire station in case additional resources were needed. Firefighter Olinik walked into the fire station, greeted other firefighters, and then suddenly collapsed.

Firefighter initiated CPR, and an ambulance was summoned. Firefighter Olinik was transported to the hospital but was pronounced dead at 0530 hours the next day. The cause of death was a heart attack.

April 3, 2007–1707 hrs
Edgar Hamlin Scott, Fire Police Captain
Age 75–Volunteer
Menands Fire Company #1, New York

At approximately 1650 hours, the Menands Fire Department, Menands Police Department, and the Town of Colonie EMS were dispatched to a vehicle crash on Interstate 787 Northbound south of Exit 6.

Firefighters found a motorcycle crash and began to provide treatment. Three Menands Fire Department fire police officers, including Fire Police Captain Scott, responded to the scene in a fire police van. Fire Police Captain Scott was driving.

As the van was completing a U-turn, it was involved in a crash with a propane tanker. Fire Police Captain Scott was ejected from the vehicle and received fatal injuries. He was not wearing a seatbelt at the time of the collision.

April 7, 2007–0106 hrs
Christopher Michael Jaros, Firefighter
Age 24–Volunteer
Ceredo Volunteer Fire-Rescue Department, West Virginia

Firefighter Jaros was responding to his fire station after his fire department was dispatched to a motor vehicle crash. Firefighter Jaros was the driver and only occupant of his personal vehicle, a 2005 Ford F–150 pickup truck.

Firefighter Jaros apparently encountered "black ice" conditions and lost control of his vehicle. His vehicle crossed the center line and collided with an oncoming vehicle, a Mazda 626. Firefighter Jaros was ejected from his vehicle and sustained fatal injuries. Firefighter Jaros was not wearing a seatbelt at the time of the crash.

April 12, 2007–Time Unknown
Bryan Zollner, North Region Chief of Staff Operations
Age 44–Career
California Department of Forestry and Fire Protection

Chief Zollner was killed in a single vehicle crash. His State-owned SUV left the roadway, went down an embankment, and struck a tree. Preliminary reports cited speed and icy roads as factors in the crash.

April 16, 2007–0630 hrs
Kyle Robert Wilson, Firefighter–Technician I
Age 24–Career
Prince William County Department of Fire and Rescue, Virginia

Technician Wilson was assigned to Tower 512, a ladder company. Tower 512 was dispatched to a reported house fire at 0603 hours. The Prince William County area was under a high wind advisory as a nor'easter moved through the area. Sustained winds of 25 miles per hour with gusts up to 48 miles per hour were prevalent in the area at the time of the fire dispatch.

Initial arriving units reported heavy fire on the exterior of two sides of the single-family house, and crews suspected that the occupants were still inside the house sleeping because of the early morning hour. A search of the upstairs bedroom was conducted by Technician Wilson and his officer. A rapid and catastrophic change of fire and smoke conditions occurred in the interior of the house within minutes of Tower 512's crew entering the structure. Technician Wilson became trapped and was unable to locate an immediate exit. "Mayday" radio transmissions of the life-threatening situation were made by crews and by Technician Wilson. Valiant and repeated rescue attempts to locate and remove Technician Wilson were made by the firefighting crews during extreme fire, heat, and smoke

conditions. Firefighters were forced from the structure as the house began to collapse on them and intense fire, heat, and smoke conditions developed. Technician Wilson succumbed to the fire and the cause of death was reported by the medical examiner to be thermal and inhalation injuries.

An extensive report on this incident is available from the Prince William Department of Fire and Rescue at http://www.pwcgov.org/ Follow the links to the Fire & Rescue main page.

For additional information regarding this incident, please refer to NIOSH Fire Fighter Fatality Investigation and Prevention Program report F2007-12 (http://www.cdc.gov/niosh/fire/reports/face200712.html).

April 25, 2007–1441 hrs
Billy Aaron Gafford, Jr., Captain–Paramedic
Age 56–Career
Birmingham Fire & Rescue Services, Alabama

Captain Gafford was participating in a work fitness evaluation with other firefighters. Captain Gafford chose the option that requires the firefighter to complete 12 laps of a college running track within 47 minutes. A paramedic unit stood by at the track, and water was available to firefighters.

Captain Gafford collapsed during lap 7. Numerous firefighters rushed to his side to provide assistance. Captain Gafford was not breathing. Firefighters began using a bag valve mask, as a defibrillator was attached. Captain Gafford was shocked, and other advanced life support treatments were provided.

Captain Gafford was transported by ambulance to the hospital. Despite efforts by firefighters and hospital personnel, he was pronounced dead in the emergency room.

April 28, 2007–0130 hrs
Joe Eddy Ivey, Captain
Age 51–Career
Nacogdoches Fire Department, Texas

Captain Ivy travelled to Houston to attend a highrise firefighting training seminar. He complained of not feeling well after dinner with other firefighters. At approximately 0130 hours, he collapsed due to a heart attack. Firefighters provided assistance and he was transported to the hospital, but he was pronounced dead at approximately 0200 hours.

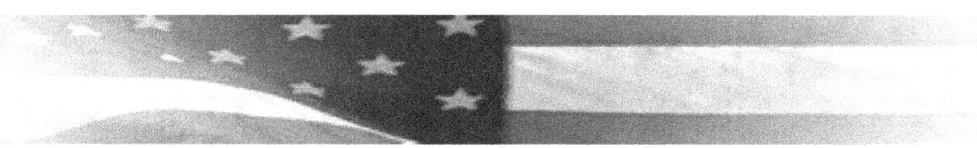

April 28, 2007–1700 hrs
Vernon Robert "Bob" McKenzie, Firefighter
Age 56–Volunteer
Gem County Fire Protection District #2, Idaho

Firefighter McKenzie completed an annual physical fitness evaluation called a "pack test." The test is required in order to work as a wildland firefighter in many jurisdictions. It requires a firefighter to walk a prescribed distance carrying a backpack of a prescribed weight within a limited time.

Firefighter McKenzie completed the test. When he arrived home, he complained of not feeling well. He was transported to a hospital, but died of a heart attack.

May 3, 2007–1230 hrs
William D. Church, Sr., Fire Police Captain
Age 63–Volunteer
Columbus Volunteer Fire Department, Pennsylvania

Fire Police Captain Church was the driver of the second-due engine to a structure fire. Prior to arriving at the scene, his unit was cancelled and returned to quarters.

Upon arrival at quarters, Fire Police Captain Church backed his engine apparatus into the fire station. He was found by other firefighters slumped over the wheel, unresponsive.

Fire Police Captain Church was treated by other firefighters and transported to the hospital. His death was the result of a heart attack.

May 6, 2007–1147 hrs
Bruce Joseph Zumwalt, Firefighter
Age 54–Volunteer
Sheldon District Fire Department, Illinois

Firefighter Zumwalt and the members of his fire department were operating at a shed fire that apparently had been caused by powerlines making contact with a structure due to high winds.

Firefighter Zumwalt was operating a handline between the burning shed and the garage to prevent fire spread into the garage. Firefighter Zumwalt operated the handline for 20 minutes or more, then collapsed without warning.

Firefighters came to his side immediately and dragged him away from the burning building. Firefighter Zumwalt was not breathing. CPR was started and an AED was attached.

continued on next page

The response time for an ambulance was going to be long, so firefighters transported Firefighter Zumwalt in a fire department vehicle. The ambulance met the firefighters during the response and a paramedic from the ambulance rode in the fire department vehicle to the hospital. Firefighter Zumwalt was not revived. His death was caused by a heart attack.

May 7, 2007–1145 hrs
Peter Beebe-Lawson, Firefighter
Age 50–Volunteer
Springfield Fire Department, Maine

Firefighter Beebe-Lawson was responding as the driver and sole occupant of a fire department tanker (tender) to a fire in a sawmill. The tanker had a water capacity of 3,500 gallons.

As he emerged from a sharp curve, Firefighter Beebe-Lawson lost control of the tanker. The right wheels of the apparatus left the roadway and ran onto the shoulder. Firefighter Beebe-Lawson overcorrected or oversteered the apparatus back onto the roadway, and it veered off the left side of the roadway, crashed into some trees, and landed on the passenger side.

Firefighter Beebe-Lawson was not wearing a seatbelt at the time of the crash. He was trapped in the wreckage, and was pronounced dead at the scene. The law enforcement report cited speed as contributing factor in the crash. Driver inexperience also was cited in press reports about the incident. The tanker was a converted heating oil tanker.

May 8, 2007–1921 hrs
Ottis Earl Stephenson, Jr., Firefighter
Age 61–Volunteer
Angier & Black River Fire Department, North Carolina

Firefighter Stephenson responded to a kitchen fire on the evening of May 8, 2007. Firefighter Stephenson responded as a part of the crew on an equipment truck. The incident was handled by the first-arriving units, and the equipment truck was released by the IC to return to quarters. The incident was concluded at approximately 2010 hours.

The next day, May 9, Firefighter Stephenson reported for work at his regular job as a security guard for a local museum. After taking a walk around the facility, Firefighter Stephenson became ill and collapsed. He was transported to the hospital, but was pronounced dead at 1236 hours. His death was caused by a heart attack.

May 11, 2007–2313 hrs
Brandon Lee Daley, Firefighter
Age 19–Volunteer
Butler County Fire District #3-Rose Hill Volunteer Fire Company, Kansas

Firefighter Daley and the members of his fire department were dispatched to a structure fire in their community. Firefighter Daley responded in his personal vehicle to the fire station. Upon his arrival at the fire station, he learned that there was no room for him to respond on the first piece of fire apparatus. Firefighter Daley was instructed to wait for the next piece of apparatus to respond.

Firefighter Daley departed the fire station in his personal vehicle, a 1990 full-size Chevrolet extended-cab pickup truck, and began his response to the fire scene. Firefighter Daley passed a car headed in the same direction but oversteered when he entered his travel lane. The right wheels of Firefighter Daley's vehicle left the right side of the paved surface of the roadway. Firefighter Daley lost control of his vehicle, it slid sideways, and rolled two-and-a-half times. Firefighter Daley was ejected from the vehicle during the second roll.

Firefighter Daley was not wearing a seatbelt at the time of the crash. He received fatal injuries, and was pronounced dead at the hospital just after midnight.

May 11, 2007–1415 hrs
Jeffrey Jeans, Firefighter–EMT
Age 46–Volunteer
Eudora Fire Department, Mississippi

Firefighter Jeans participated in a large-diameter hose drill on the evening of May 10. He spent the night at the fire station. On the morning of May 11, Firefighter Jeans retrieved a fire truck from a repair shop and brought the truck back to the fire station.

Once at the fire station, Firefighter Jeans singlehandedly reloaded the large-diameter hose on the repaired apparatus. He also performed a number of other maintenance tasks around the fire station in preparation for training to be held that weekend.

At approximately 1415 hours, Firefighter Jeans went into the fire station and told other firefighters that he was not feeling well. He went into the kitchen area of the station and suddenly collapsed. Firefighters provided aid immediately, and Firefighter Jeans was transported to the hospital. Despite the efforts of firefighters and hospital personnel, Firefighter Jeans did not recover from his heart attack.

May 14, 2007–1004 hrs
Edward Levirgil Andrews IV, Fire Captain
Age 53–Career
Redding Fire Department, California

Captain Andrews reported for duty at approximately 0750 hours on May 14, 2007. Captain Andrews and his partner prepared for the day by completing station and apparatus maintenance duties.

Between 0830 hours and 0900 hours, Captain Andrews and another crew member drove to an area of their district that contained walking trails. The firefighters walked for some distance and then began to walk up a steep roadway. Captain Andrews complained of heartburn but dismissed it.

The other firefighter began to run as Captain Andrews continued to walk. After a short time apart, the firefighter saw Captain Andrews on his back in apparent distress. The firefighter called for assistance and found Captain Andrews unresponsive.

Other firefighters and EMS workers arrived and transported Captain Andrews to the hospital. Captain Andrews was pronounced dead by the emergency room physician.

May 16, 2007–1828 hrs
Joseph E. Piazzi, Deputy Chief and Fire Police Captain
Age 76–Volunteer
Briarcliff Manor Fire Department, New York

Deputy Chief Piazzi had responded to several incidents of downed powerlines and trees due to storms throughout the afternoon of May 16, 2007. He then returned home and assisted a neighbor with cleanup.

His fire department was dispatched to another call and he was preparing to respond when he began to experience chest pains. Deputy Chief Piazzi contacted 9-1-1 and the fire department was dispatched to his residence. Upon their arrival, he was found unresponsive. Care was initiated, and he was transported to the hospital, where he was pronounced dead.

May 19, 2007–1033 hrs
John Francis Keane, Captain
Age 37–Career
Waterbury Fire Department, Connecticut

Captain Keane was the Company Officer for Engine 8. At 1034 hours on May 22, 2007, Engine 8 and several other fire department units were dispatched to a possible kitchen fire in an apartment. Engine 8 was operating an older reserve piece of fire apparatus, since their apparatus was in the

shop. In addition, Engine 8 was responding from a location other than their fire station. The crew was attending a blood drive.

Engine 8 and Truck 1 entered an intersection at the same time from different directions. Truck 1 struck the left front of Engine 8. The impact caused the engine to spin clockwise approximately 180 degrees. Captain Keane and the driver of Engine 8 were ejected from the apparatus. The officer on Truck 1 was trapped in the vehicle and had to be extricated. All eight firefighters from Engine 8 and Truck 1 were injured.

Captain Keane was transported to the hospital, where he remained until his death on May 22, 2007. The cause of death was listed as blunt head trauma.

A law enforcement investigation of the crash stated that Truck 1 had the green light, that a number of firefighters, including Captain Keane, were not wearing seatbelts at the time of the crash, and that some lines of sight at the crash scene were obstructed by trees.

May 20, 2007–1445 hrs
Dennis Cheshire, Firefighter
Age 48–Volunteer
Red Oak Area Volunteer Fire Department, Alabama

Firefighter Cheshire was operating a military surplus tractor-trailer tanker (tender) responding to a structure fire. During the response, Firefighter Cheshire made a sharp left-hand turn into a T-intersection.

The apparatus made a turn that was too wide, ran off of the right shoulder of the road, and then rolled down an embankment. Firefighter Cheshire was pinned under the cab of the vehicle and was pronounced dead at the scene.

The vehicle was not equipped with seatbelts. The tanker's speed at the time of the crash was estimated at 5 miles per hour.

May 25, 2007–1530 hrs
Bradley William "Wally" Green, Firefighter
Age 53–Volunteer
Monroe Township/Cowan Volunteer Fire Department, Indiana

Firefighter Green and the members of his fire department worked for over 4 hours at the scene of a rolled over fertilizer tanker. Firefighter Green was on the scene for the entire incident and staffed a precautionary handline.

continued on next page

Firefighter Green complained of being overheated. He removed his protective clothing and cooled off with water under a tree. When the incident was concluded, at approximately 1148 hours, Firefighter Green left the scene.

At approximately 1530 hours, Firefighter Green collapsed due to an apparent heart attack. A bystander performed CPR until emergency response personnel arrived, but Firefighter Green did not survive.

May 28, 2007–0503 hrs
Felix Maurice Roberts, Firefighter II
Age 41–Career
Fulton County Fire Department, Georgia

Firefighter Roberts and other firefighters were dispatched to a residential structure fire. Upon their arrival, firefighters found a working fire with first-hand reports of a civilian trapped in the structure. Firefighter Roberts and another firefighter advanced a hoseline into the structure for search and rescue and fire control.

Fire conditions inside the structure changed rapidly, and most firefighters were forced to evacuate the structure. Firefighter Roberts and his partner were trapped in the structure until they were located and removed by other firefighters.

Firefighter Roberts was not breathing when he was removed from the structure. He was transported by ambulance to a medical facility, but was not revived. The cause of death was listed as asphyxiation. Firefighter Roberts' carboxyhemoglobin level was 14 percent.

May 28, 2007–1700 hrs
Henry Lawrence "Hank" Pitts, Jr., Lieutenant
Age 56–Career
Douglas County Fire/EMS, Georgia

Lieutenant Pitts worked a double shift from approximately 0800 hours on May 26 through approximately 0800 hours on May 28. During this period, he responded to at least seven emergency incidents, including fires and EMS incidents.

After going off duty, Lieutenant Pitts and his wife visited with friends at a cabin in North Georgia. At approximately 1700 hours on May 28, Lieutenant Pitts was found, unresponsive, near his friend's cabin.

continued on next page

Lieutenant Pitts was transported by medical helicopter to a regional medical facility. He underwent surgery but did not recover. Lieutenant Pitts died as the result of a heart attack on June 1, 2007.

May 28, 2007–Time Unknown
David A. Middleton, Firefighter
Age 38–Career
Boston Fire Department, Massachusetts

Firefighter Middleton was on duty with his engine company the night of May 27-28, 2007. During the shift, the crew had responded to several emergencies, including a working structure fire.

When his shift ended, Firefighter Middleton told other firefighters that he was not feeling well and skipped his normal workout. Firefighter Middleton went home and suffered a fatal heart attack shortly after arriving at his residence.

June 4, 2007–1343 hrs
Mark Stevens Carter, Engineer–Paramedic
Age 53–Career
Phoenix Fire Department, Arizona

Engineer Carter was assigned to Engine 37. He reported for duty at approximately 0800 hours and performed apparatus and fire station maintenance duties. His crew met the crew of the ladder company from the same station at a local high school to participate in department-mandated physical fitness activities.

The crews played basketball all morning with three emergency responses mixed in, all EMS incidents. At approximately 1330 hours, Engineer Carter and his crew stopped at a restaurant for lunch. Engineer Carter placed his order, paid for the meal, and went outside. As he walked toward his apparatus, Engineer Carter suddenly collapsed.

Bystanders summoned firefighters from the restaurant, and paramedic-level treatment began immediately. Engineer Carter was transported to the hospital but was not revived. The cause of the death was listed as a heart attack.

June 12, 2007–2000 hrs
David Allan Rufer, Firefighter
Age 42–Paid-on-Call
Monroe Fire Department, Wisconsin

Firefighter Rufer and other firefighters were engaged in a training exercise that involved the use of SCBA and fire hose. During the course of the exercise, Firefighter Rufer collapsed due to an apparent heart attack.

Firefighter Rufer was immediately treated by other firefighters and transported to the hospital, but he did not survive.

June 18, 2007–1930 hrs
Theodore Michael Benke, Captain
Age 49–Career
Charleston Fire Department, South Carolina

William H. "Billy" Hutchinson, Captain
Age 48–Career
Charleston Fire Department, South Carolina

Louis Mark Mulkey, Captain
Age 34–Career
Charleston Fire Department, South Carolina

Mark Wesley Kelsey, Acting Captain
Age 40–Career
Charleston Fire Department, South Carolina

Bradford Rodney "Brad" Baity, Engineer
Age 37–Career
Charleston Fire Department, South Carolina

Michael Jonathon Alan French, Assistant Engineer
Age 27–Career
Charleston Fire Department, South Carolina

Melvin Edward Champaign, Firefighter
Age 46–Career
Charleston Fire Department, South Carolina

James Allen "Earl" Drayton, Firefighter
Age 56–Career
Charleston Fire Department, South Carolina

Brandon Kenyon Thompson, Firefighter
Age 27–Career
Charleston Fire Department, South Carolina

At 1909 hours, the Charleston Fire Department received a report of a fire at the Sofa Super Store, 1807 Savannah Highway, in Charleston. Firefighters arriving on the scene found an exterior fire extending into the interior of the commercial structure. Firefighters at the rear of the structure extended hoselines and fought the exterior and interior fires at that location.

Other firefighters entered the front of the structure to search for building occupants and to fight the fire in the front retail area of the store; they were wearing full structural firefighting protective clothing and SCBA. While initial conditions inside the retail area of the store were tenable, conditions quickly worsened as the fire extended from the rear of the store into the retail area of the store.

Approximately 20 minutes after arrival on the scene, firefighters in distress could be heard on the radio requesting assistance. Smoke conditions in the interior of the store had worsened significantly, and firefighters were unable to find their way to a safe exit. The fire progressed rapidly and involved the interior of the store completely.

Nine firefighters were trapped in the structure and were unable to escape. Their bodies were recovered and removed from the structure later that night after the main body of fire was extinguished. The Charleston County Coroner's Office took possession of the bodies. The cause of death for all nine firefighters was attributed to smoke inhalation and burns.

An extensive report on this incident is available from the City of Charleston at http://www. charlestoncity.info/ Follow the links to the fire department main page.

June 21, 2007–1654 hrs
Daniel F. Pujdak, Firefighter
Age 23–Career
Fire Department City of New York, New York

Firefighter Pujdak was a member of the crew of Ladder 146 based in Brooklyn. Firefighter Pujdak and his crew were dispatched, along with other firefighters, to a fire in a four-story building.

As Firefighter Pujdak stepped from his ladder truck's aerial ladder to the roof of the structure, he slipped and fell to the ground. Firefighter Pujdak received fatal injuries.

June 22, 2007–1525 hrs
Timothy Lavern Sanborn, Firefighter
Age 56–Volunteer
Clinton Area Fire and Rescue, Michigan

Firefighter Sanborn was operating the pump, supplying water to multiple attack lines on a residential structure fire. Approximately 30 minutes into the incident, as the fire was controlled, Firefighter Sanborn told other firefighters that he was not feeling well.

Firefighter Sanborn walked under his own power to an ambulance on the scene. EMS personnel treated Firefighter Sanborn for his heart attack symptoms. The ambulance departed for the hospital. While en route to the hospital, Firefighter Sanborn became unconscious. Despite treatment in the ambulance and in the hospital, Firefighter Sanborn was pronounced dead at 1620 hours.

June 28, 2007–1020 hrs
Edward Charles Summers, Firefighter
Age 69–Volunteer
Patchogue Fire Department, New York

Firefighter Summers was participating in a work detail at the fire station when he collapsed of a heart attack. Firefighters provided emergency care immediately but Firefighter Summers could not be revived.

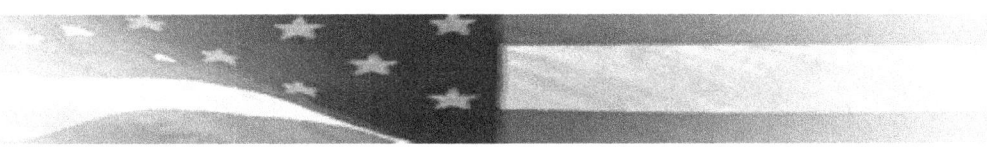

June 30, 2007–0930 hrs
Samuel W. Downing, Captain
Age 45–Career
Mobile Fire-Rescue Department, Alabama

Captain Downing had worked a 24-hour shift ending at approximately 0730 hours on June 30, 2007. During the shift, Captain Downing and his crew had responded to multiple emergency incidents, including a vehicle crash that occurred just before shift change.

Captain Downing departed his fire station and went home. Shortly after arriving home, Captain Downing experienced a fatal heart attack. He was pronounced dead at approximately 0930 hours.

July 2, 2007–0400 hrs
Michael B. Douthitt, Driver–Engineer
Age 48–Career
Broward County Sheriff's Office Department of Fire/Rescue, Florida

Driver-Engineer Douthitt and other firefighters operated at the scene of a residential structure fire at approximately 0400 hours on July 2, 2007.

Driver-Engineer Douthitt participated in overhaul activities in the residence. Upon returning to the fire station after the response, Driver-Engineer Douthitt felt ill. He rested at home for 2 days and the symptoms went away; he returned to work on July 4, 2007.

After going home on July 5, 2007, Driver-Engineer Douthitt was hospitalized for a cardiac event. He underwent a heart catheterization and was released from the hospital.

He completed paperwork to document his illness on July 8, 2007. On July 13, 2007, while at home, Driver-Engineer Douthitt went into cardiac arrest. He later died at a local hospital.

July 2, 2007–1630 hrs
Ronald Yale Wiley, Deputy Fire Marshal
Age 47–Career
Richmond Fire Department, California

Deputy Fire Marshal Wiley was returning in his city vehicle to his office from a meeting. As he drove over the Carquinez Bridge, he was involved in a single vehicle crash and fire. Deputy Fire Marshal Wiley died in the crash.

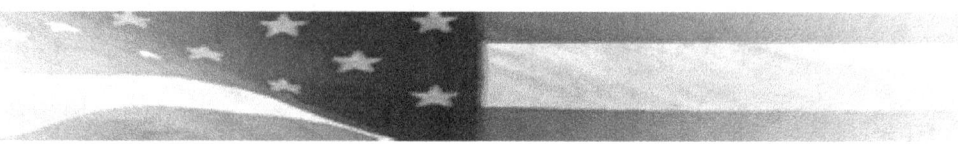

July 2, 2007–1640 hrs
Dennise Marie Leslie, Firefighter
Age 40–Volunteer
Coal City Community Volunteer Fire Department, Indiana

Firefighter Leslie was responding to the fire station in her personal vehicle, a 2000 Ford Ranger pickup truck, after her fire department was dispatched to a wildland fire.

As Firefighter Leslie crested a hill, she saw another vehicle waiting to make a left-hand turn in front of her. She moved her vehicle from the center of the road into her lane, but oversteered. The right wheels of her vehicle left the paved roadway. Firefighter Leslie steered to the left in an attempt to bring her vehicle back onto the road. The pickup began to spin, travelled across the roadway, and collided with an earthen embankment. This caused the vehicle to go airborne and begin to roll over. The cab of the truck collided with a large tree, causing extensive damage to the passenger compartment.

When EMS responders arrived on the scene, Firefighter Leslie was determined to have died in the crash. The cause of death was listed as head trauma. The pickup's airbag had deployed; Firefighter Leslie was not wearing a seatbelt.

July 5, 2007–0014 hrs
Eric Robert Lyons, Firefighter
Age 37–Career
Kennewick Fire Department, Washington

Firefighter Lyons responded to a wildland fire during the evening hours of July 4. He and the rest of his crew returned to their fire station after the conclusion of the incident.

Shortly after midnight on July 5, 2007, Firefighter Lyons' coworkers heard him making snoring-type sounds in the bunkroom. Firefighters checked on Firefighter Lyons' welfare and found that he was experiencing a medical emergency.

Firefighters provided medical assistance, and Firefighter Lyons was transported to the hospital but later died. The cause of death was cardiac-related.

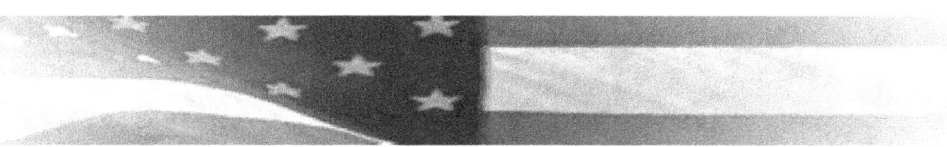

July 5, 2007–1000 hrs
Michael James Penovich, Fire Chief
Age 38–Career
Saratoga Springs Fire Department, Utah

Chief Penovich was returning to his community after going to Heber City to look at a water tank that his department was considering for purchase. Chief Penovich was driving his department-issued F-150 Ford pickup truck.

For unknown reasons, Chief Penovich's vehicle veered off of the roadway, over an embankment, and into the Deer Creek Reservoir. Witnesses stated that the vehicle did not slow down as it left the roadway. The vehicle submerged immediately. Bystanders on the roadway and in boats attempted to provide aid but could not reach the vehicle.

The Wasatch County Search and Rescue divers entered the water at 1048 hours and surfaced at 1055 hours. They reported that they had located Chief Penovich in his vehicle. They found his seatbelt across his chest but not latched. Chief Penovich was halfway out of the passenger side window. An autopsy showed that he had drowned.

July 6, 2007–0050 hrs
James J. McRae III, Firefighter–Technician
Age 34–Career
District of Columbia Fire and Emergency Medical Services

Firefighter McRae was on duty as the driver of Truck 12, having reported for duty at 0700 hours on July 6, 2007. During the course of the shift, Firefighter McRae and the members of his fire company performed fire station and apparatus maintenance duties, conducted a fire inspection, attended a meeting, and responded to three emergency incidents.

At 0050 hours on July 7, 2007, Firefighter McRae awakened another paramedic firefighter and asked the firefighter to medically assess him. Firefighter McRae was not feeling well, he was having difficulty breathing, and he was coughing. Three minutes later, an ambulance was requested.

Firefighters provided paramedic-level emergency medical care to Firefighter McRae, and he was transported to a hospital by ambulance. At approximately 0315 hours, Firefighter McRae was transferred by medical helicopter to another hospital, where he was pronounced dead at 0415 hours. The cause of death was a heart attack.

For additional information regarding this incident, please refer to NIOSH Fire Fighter Fatality Investigation and Prevention Program report F2007-23 (http://www.cdc.gov/niosh/fire/reports/face200723.html).

July 15, 2007–1300 hrs
Stephen R. Dembski, Firefighter
Age 41–Volunteer
Fire Department Ridgefield Park, New Jersey

Firefighter Dembski was at home at 0658 hours when his fire department was paged to respond to a mutual-aid residential structure fire. Firefighter Dembski responded to the fire station in his personal vehicle, and then to the fire scene aboard his fire department's ladder apparatus.

Upon arrival at the scene, Firefighter Dembski performed forcible entry and search and rescue duties on the second floor of the structure. When the incident was concluded, Firefighter Dembski and other firefighters restored their equipment and apparatus to service. Firefighter Dembski left the fire station at approximately 1110 hours.

Firefighter Dembski attended a child's birthday party that afternoon. At approximately 1500 hours, Firefighter Dembski collapsed due to an apparent heart attack. Firefighters at the party provided medical assistance, and an ambulance and paramedics were summoned. Firefighter Dembski was transported to the hospital but was pronounced dead at 1548 hours. The cause of death was listed as a ruptured aorta.

July 21, 2007–0200 hrs
Matthew Charles Burton, Captain
Age 34–Career
Contra Costa County Fire Protection District, California

Scott Peter Desmond, Fire Engineer
Age 37–Career
Contra Costa County Fire Protection District, California

Captain Burton, Engineer Desmond, and another engineer were the crew of Engine 70. At 0143 hours, Engine 70 was dispatched to a residential fire alarm. As additional information was received, the incident was upgraded to a structural fire response with the addition of two engines, a quint, and a command officer.

Engine 70 arrived on the scene at 0150 hours and reported heavy fire and smoke from a small single-family residence. Firefighters reported that they had confirmed reports that two occupants of the home were still inside.

Captain Burton and Engineer Desmond advanced an attack line into the structure and flowed water on the fire. They reported that the fire had been knocked down and requested ventilation at 0155 hours. Captain Burton and Engineer Desmond exited the structure temporarily to retrieve a TIC, then re-entered the structure and went to the left toward the bedrooms with an attack line, while another crew went to the right without an attack line.

continued on next page

The engineer for Engine 70 placed a PPV fan at the front door. One of the civilian fire victims was located by the crew that had gone to the right; her removal was difficult and firefighters had to exit the building to ask for help. During this time, the fire inside the house advanced rapidly.

Firefighters had difficulty venting the roof due to multiple roofs, built-up roofing materials, and the type of construction.

A command officer arrived on the scene at approximately 0202 hours and began to look for Captain Burton to assume Command. The command officer tried to contact the Engine 70 crew by radio but was unsuccessful. A second alarm was requested, and a report of a missing firefighter was transmitted at approximately 0205 hours.

The fire had advanced within the structure and had to be controlled before firefighters could search for the missing crew. Captain Burton and Engineer Desmond were located and removed from the structure between 0212 and 0226 hours. The firefighters were found in a bedroom.

It has been suggested that the deceased firefighters left the handline that they advanced into the structure to conduct a search. When fire conditions changed rapidly, they were trapped in the bedroom. The cause of death for both firefighters was listed as thermal burns and smoke inhalation. The post mortem carboxyhemoglobin level for Captain Burton was 23.5 percent, and the level for Engineer Desmond was 19 percent. The cause of the fire was careless disposal of smoking materials.

July 23, 2007–0429 hrs
Cornelius Myron "Storm" Nolton, Firefighter
Age 37–Career
Newark Fire Department, New Jersey

Firefighter Nolton was fighting a structural fire on July 23, 2007, when he began to complain of back and neck pain. He was transported from the fireground to the hospital. He was evaluated and released from the hospital by 1400 hours.

The pain continued and Firefighter Nolton went to a different hospital at approximately 1730 hours. He remained in the hospital until his death on July 25, 2007. His death was caused by a heart attack.

July 23, 2007–1015 hrs
Dennis Luster Davis, Pilot
Age 61–Wildland Contract
Idaho Helicopters under contract to USDA Forest Service, Klamath National Forest, California

Pilot Davis was dropping a package of drinking water for firefighting crews engaged in the Elk Complex of wildland fires. The water was suspended from the underside of the helicopter by a 150-foot line.

Firefighters on the ground watched as the package of water was set down on the ground. The helicopter drifted to the right and the rotor contacted a tree. The helicopter crashed, and a postcrash fire consumed most of the aircraft. Pilot Davis was killed in the crash.

For additional information about this crash, consult the National Transportation Safety Board Web site at http://www.ntsb.gov/ntsb/query.asp - NTSB identification LAX07TA227.

July 24, 2007–2107 hrs
Jon Charles Trainer, Firefighter
Age 38–Volunteer
Mechanicsburg Fire Department, Ohio

Firefighter Trainer and other firefighters were cleaning and restoring fire apparatus to service after fighting a residential structure fire. Firefighter Trainer was on top of an engine apparatus assisting with hose loading. He slipped and fell from the top of the apparatus and sustained a severe head injury.

Firefighter Trainer was transported to the hospital by ambulance but died on July 25, 2007.

July 27, 2007–0415 hrs
James Glenn "Shib" Miller, Firefighter
Age 43–Volunteer
Sesser Fire Protection District, Illinois

The Sesser Fire Department responded to a fire involving a tractor-trailer truck on a local Interstate. Three fire trucks were parked on scene on the right shoulder and in the first traffic lane to the left of the shoulder. Safety cones were in place on the roadway, and the emergency lights on the fire apparatus were activated.

continued on next page

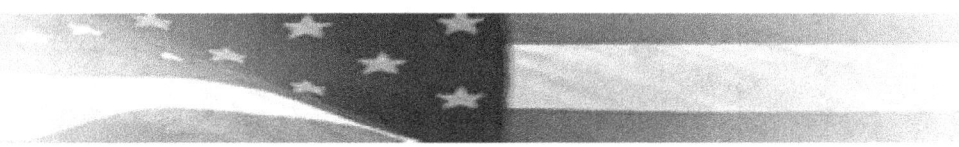

The response of a State law enforcement officer was delayed. The fire department IC declined offers of assistance from local law enforcement for traffic control, citing the lack of traffic on the highway.

Firefighter Miller was replacing some equipment that had been used at the incident scene into a driver's side compartment on one of the fire trucks. At approximately 0415 hours, Firefighter Miller was struck by a passing bus. He was thrown over 200 feet to the side of the road. He suffered fatal injuries.

Firefighters on the scene provided emergency medical treatment to no avail. The driver of the bus was charged with reckless homicide and was ordered to stand trial.

July 27, 2007–1251 hrs
Nemeth Fitzhugh Sanders, Forest Fire Equipment Operator
Age 43–Wildland Full-Time
North Carolina Division of Forest Resources

Forest Fire Equipment Operator Sanders was returning home after completing training. He was driving a State-owned vehicle on Interstate 40 near Morrisville.

A traffic collision in the opposing direction of travel caused a tractor-trailer truck to cross through the median into Forest Fire Equipment Operator Sanders' lane. His vehicle struck and went partially underneath the trailer. The trailer, his vehicle, and another car that struck the trailer burst into flames.

Forest Fire Equipment Operator Sanders died of traumatic injuries.

August 3, 2007–0138 hrs
Kevin Glenn Williams, Captain
Age 42–Volunteer
Noonday Volunteer Fire Department, Texas

Austin Hague Cheek, Firefighter
Age 19–Volunteer
Noonday Volunteer Fire Department, Texas

Captain Williams and Firefighter Cheek responded to a mutual-aid structure fire in a residence. When they arrived on the scene, the firefighters donned full protective clothing and entered the structure with a handline.

continued on next page

The fire appeared to be under control, and theirs was the third line to enter the structure. Captain Williams communicated with the IC by radio, asking about the status of electrical service to the building. After not hearing from Captain Williams or Firefighter Cheek for approximately 15 minutes, a RIC was sent in to find the two firefighters.

Both firefighters were removed from the building. Both firefighters died of burns. The details of the incident are unknown, but factors that may have played a part in the incident were a PPV fan placed at the door and renovations to the house that created a double roof.

August 8, 2007–0800 hrs
Gerald David "Jerry" Donley, Jr., Deputy Fire Chief
Age 45–Career
Swepsonville Volunteer Fire Department, Inc., North Carolina

Deputy Chief Donley responded with other members of his fire department to a mutual-aid structure fire on August 6, 2007. The incident was dispatched at 1633 hours and firefighters were back in quarters at 2149 hours. Deputy Chief Donley remained at the fire station until 0005 hours on August 8, 2007.

Deputy Chief Donley arrived at the fire department at approximately 0800 hours the morning of August 8 to work his scheduled 24-hour shift. He spent the day cleaning up equipment from the fire call that had occurred the previous day and responding to at least two EMS incidents. Later during his shift, Deputy Chief Donley went to bed at the station.

The next morning, firefighters found that Deputy Chief Donley had died during the night. The cause of death was a heart attack.

August 10, 2007–0149 hrs
Paul Darrin Baker, Lieutenant
Age 39–Volunteer
Turtle Creek Volunteer Fire Department, Arkansas

A vehicle crash occurred on a local Interstate highway. One of the vehicles involved in the crash ended up parked on the shoulder of the Saline River Bridge. Firefighters were dispatched to the incident to control a gasoline leak.

Lieutenant Baker and other firefighters arrived on the scene and controlled the gasoline leak. Absorbent material was poured on the leaked fuel.

continued on next page

Lieutenant Baker mounted the hood of the vehicle and attempted to reach in through the driver's window to shut of the ignition. Lieutenant Baker slipped and fell approximately 40 feet off of the bridge to the ground. Lieutenant Baker was pronounced dead at the scene.

August 11, 2007–1626 hrs
Todd Whitney Hage, Firefighter
Age 42–Volunteer
Wesley Chapel Volunteer Fire Department, North Carolina

Firefighter Hage was the driver and sole occupant of a commercial chassis pumper that was responding to an automatic fire alarm at a school.

As Firefighter Hage responded, he swerved to avoid an oncoming vehicle. The pumper went off the roadway, then rolled and slid into a tree. The vehicle sustained major damage, and Firefighter Hage was killed in the crash.

The cause of death was listed as blunt head and chest trauma. Firefighter Hage was wearing his seatbelt at the time of the crash.

August 13, 2007–1820 hrs
Anthony Philip Cox, Captain
Age 44–Career
Topeka Fire Department, Kansas

Captain Cox and other firefighters were working on the scene of a two-story apartment building fire. Captain Cox collapsed after leaving the building to rehab. His death was caused by a heart attack.

August 14, 2007–1000 hrs
Glenn Williams Miller, Probationary Firefighter
Age 34–Volunteer
Whispering Pines Volunteer Fire Department, North Carolina

Firefighter Miller was participating in firefighter training through a local community college. Firefighter Miller became ill during training and was transported to the hospital.

While in the hospital, Firefighter Miller underwent surgery for a heart illness but did not recover from the procedure. Firefighter Miller died on August 17, 2007.

August 14, 2007–2333 hrs
Michael Paul Stephen Heuer, Fire Chief
Age 55–Volunteer
Sierra City Fire Department, California

Chief Heuer and the members of his fire department were engaged in a protracted rescue incident. A fisherman had fractured his ankle in a creek area that was very difficult to access. The fisherman had to be carried by hand over very rough and steep terrain. The incident began at 2025 hours and the fisherman was removed to a waiting ambulance at 2312 hours.

As he helped to gather equipment used in the rescue operation, Chief Heuer became ill and sat down. Other firefighters came to his aid and assessed his condition. CPR was begun immediately and a call for a firefighter down was transmitted at 2334 hours.

Firefighters carried Chief Heuer to the roadway and to a helicopter landing zone by ambulance. Numerous shocks were given by an AED. When the medical helicopter arrived, at approximately 0020 hours, Chief Heuer's condition was once again assessed. Due to the length of time he had been ill and the lack of response to treatment, he was pronounced dead. The cause of death was listed as a heart attack.

August 18, 2007–1500 hrs
Robert C. Beddia, Firefighter
Age 53–Career
Fire Department City of New York

Joseph Graffagnino, Firefighter
Age 32–Career
Fire Department City of New York

Firefighter Beddia and Firefighter Graffagnino were assigned to the Engine 24/Ladder 5 fire station in the SoHo section of Manhattan. Both firefighters were assigned to Engine 24.

Firefighters were dispatched to a highrise office building at the former site of the World Trade Center. The building had been damaged on September 11, 2001, and was in the process of being demolished. The exterior of the building was covered with netting, and the interior was a maze of temporary containment walls and partially demolished structure. Among a number of challenges for firefighters, the building's standpipe system was partially dismantled and did not function properly.

A number of firefighters became disoriented in the building. Firefighter Beddia and Firefighter Graffagnino were located by other firefighters, removed from the building, and transported to hospitals. The preliminary cause of death for both firefighters was asphyxiation.

The Fire Department City of New York continues to investigate the incident and the conditions that led to the death of the two firefighters.

August 19, 2007–1900 hrs
Jeffrey Dale Swartz, Firefighter
Age 36–Volunteer
Wagener Fire Department, South Carolina

Firefighter Swartz was responding to a medical emergency in his personal vehicle. He lost control of his vehicle, and it left the roadway. The vehicle rolled and burst into flames when it came to rest.

Firefighter Swartz died from asphyxiation and thermal injuries. He was wearing his seatbelt at the time of the crash.

August 24, 2007–1330 hrs
M.L. Hopper, Forestry Aide 2
Age 74–Wildland Full-Time
Tennessee Department of Agriculture, Division of Forestry

Forestry Aide Hopper was flagging a proposed fire control line in the Natchez Trace State Forest near Lexington. As he worked, Forestry Aide Hopper suffered a fatal heart attack.

August 29, 2007–2114 hrs
Paul J. Cahill, Firefighter
Age 55–Career
Boston Fire Department, Massachusetts

Warren J. Payne, Firefighter
Age 53–Career
Boston Fire Department, Massachusetts

Box 281 was transmitted at 2106 hours for a building fire at an Asian restaurant in the West Roxbury section of Boston. Engine Company 30 and Ladder Company 25 responded to the alarm from quarters and arrived at the fire location at 2108 hours. Ladder Company 25, the first company to arrive, reported fire showing from the roof.

Initially, the interior of the restaurant was clear. Firefighters discovered fire in the void space above the kitchen area and began to apply water to the fire from a handline. Fire conditions changed rapidly. The Boston Fire Department Board of Inquiry report stated:

The Board of Inquiry believes oxygen entrained within the fire stream together with the infusion of oxygen from the air below the ceiling admitted a fresh supply of oxygen into the ceiling area directly above the kitchen's hood/exhaust duct and into the ceiling void space created by the dislodged

continued on next page

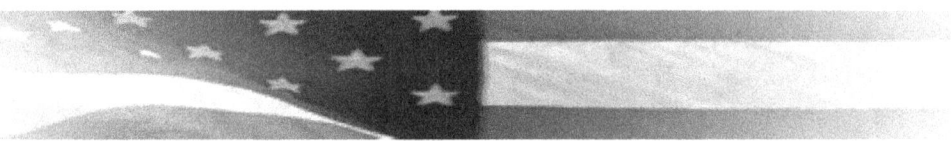

ceiling tiles. This event in turn, caused the unburned flammable gas generated by the undetected extended burning fire at that location to ignite.

Firefighter Cahill was unable to exit the kitchen area and died of asphyxiation. Firefighter Payne died of smoke inhalation and thermal injuries. Firefighter Payne was found in the seating area of the restaurant.

September 8, 2007–0914 hrs
George H. Crotts, Jr., Engineer
Age 70–Volunteer
Willow Grove Volunteer Fire Company, Pennsylvania

Engineer Crotts sustained a serious head injury when he fell from the top of an antique fire engine that was being loaded on a trailer for transport to a parade. Engineer Crotts was transported to the hospital but died of his injuries on September 9, 2007.

September 9, 2007–1150 hrs
Jared W. Zimmerly, Firefighter
Age 20–Volunteer
Prairie Township Fire Department, Ohio

Firefighter Zimmerly was responding to a mutual-aid structure fire in his personal vehicle, a 2006 Pontiac GTO.

Firefighter Zimmerly lost control of his vehicle in a slight left-hand curve in the road. His vehicle left the left side of the roadway, crashed through a sign and several trees, and came to rest on its roof. Firefighter Zimmerly was not wearing his seatbelt at the time of the crash. He was ejected from the vehicle and ended up underneath the car. He was pronounced dead at the scene.

September 10, 2007–1930 hrs
Michael Dean Stanfield, Captain
Age 30–Volunteer
Anderson Township Volunteer Fire Department, North Carolina

Captain Stanfield responded to the scene of a motor vehicle crash in his personal vehicle. There was no crash scene located, so firefighters were released from the incident scene to return to quarters.

Captain Stanfield and another firefighter exchanged personal vehicles at the scene and planned to drive to the fire station in each other's vehicles. Captain Stanfield was operating a 2004 Honda motorcycle. He was wearing a helmet.

Captain Stanfield was involved in a single vehicle crash while driving to the fire station. The motorcycle left the left side of the road and struck a ditch. The motorcycle overturned, and Captain Stanfield was fatally injured.

Firefighters responded to the scene and provided emergency medical treatment. Despite aggressive treatment and rapid transport to the hospital, Captain Stanfield did not survive.

September 12, 2007–0115 hrs
Leonard R. Bailey, Jr., Fire Chief
Age 56–Volunteer
Elizabeth Volunteer Fire Department, Pennsylvania

Chief Bailey was at home when he was notified that a building near his home was on fire. The building was a bar that had previously been owned by Chief Bailey's father.

Chief Bailey left his home to investigate the report and found that there was smoke showing from the structure. Chief Bailey ran back to his residence, reported the fire and his findings to the fire department, and then suffered a heart attack.

September 14, 2007–0420 hrs
Jerry Wayne Stucker, Firefighter–EMT
Age 53–Industrial
Dow Corning Fire Department-Loss Prevention Department, Indiana

Firefighter Stucker responded to a hazardous materials emergency. He was attaching a hoseline to a fire hydrant when he suffered a fatal heart attack.

September 14, 2007–2015 hrs
Kelly L. Page, Firefighter
Age 38–Career
City of Lowell Fire Department, Massachusetts

Firefighter Page had worked a number of calls during the day including downed powerlines, medical emergencies, false alarms, and a basement fire in a vacant residential structure. In the evening after the last of the calls, Firefighter Page complained to other firefighters about not feeling well.

A short time later Firefighter Page was found unconscious on the floor of the station. Firefighters attempted to resuscitate him; he was transported to Saints Medical Center where he was pronounced dead due to a heart attack.

September 15, 2007–1558 hrs
William D. "Billy" McDaniels, Sr., Fire Police Captain
Age 51–Volunteer
Mocanaqua Volunteer Fire Company Number 1, Pennsylvania

Fire Police Captain McDaniels responded with other members of his fire department to an alarm in a local highrise. When Fire Police Captain McDaniels arrived on the scene, he told other firefighters that he was not feeling well. He was immediately treated and transported to the hospital. He was pronounced dead due to a heart attack at 1737 hours.

September 24, 2007–1414 hrs
Bryon Wayne Johnson, Lieutenant
Age 32–Career
Sedgwick County Fire District #1, Kansas

The fire department communications center received several calls reporting a brush fire that may have been caused by downed powerlines. Squad 34, under the command of Lieutenant Johnson, responded to the call but did not acknowledge the message about powerlines.

When the unit arrived on the scene, powerlines were visible in the street. Firefighters advanced a handline and began to control the fire. In order to access a portion of the fire, the apparatus had to be backed up. Lieutenant Johnson advanced the hoseline, and his engineer drove the apparatus. The two firefighters communicated by hand signal.

Lieutenant Johnson was hitting hot spots with the handline. His engineer viewed him in the rearview mirror. The engineer saw Lieutenant Johnson stiffen and fall. The engineer stopped the apparatus, got out, and found Lieutenant Johnson on his back.

continued on next page

The engineer recognized the presence of an energized powerline. With some difficulty, the engineer pulled Lieutenant Johnson away from the hazard. The engineer called to another firefighter who had just arrived on the scene for help. The firefighter reported the incident, ordered the dispatch of an ambulance, and requested a command officer and additional firefighters.

Lieutenant Johnson was transported to the hospital but did not survive. The cause of death was listed as electrocution. The fallen powerline that caused the grass fire was the result of a pole being struck by a truck making a U-turn.

September 25, 2007–2000 hrs
John E. "Dawg" Lietzke, Firefighter
Age 47–Volunteer
Olivet Fire Department, Michigan

Firefighter Lietzke was responding in his personal vehicle to an emergency. He struck a large tree that had fallen in the road. He was pronounced dead at the scene. Firefighter Lietzke was wearing his seatbelt at the time of the crash.

September 26, 2007–2306 hrs
Mike D. Reagan, Jr., Firefighter
Age 19–Volunteer
Sharon Hill Fire Company, Pennsylvania

Firefighter Reagan and other firefighters were operating at a fire in a detached garage. The fire started when a home owner accidentally spilled gasoline while performing repairs on a motorcycle. The fire had been knocked down and firefighters were overhauling and putting out hot spots.

A sudden structural collapsed occurred and trapped Firefighter Reagan. At least two other firefighters were trapped by the collapse. Other firefighters freed the trapped firefighters, and Firefighter Reagan was transported to the hospital for treatment. Firefighter Reagan did not recover from his injuries; he died on September 29.

October 4, 2007–1956 hrs
Adam Edward Cole, Firefighter
Age 24–Volunteer
Buchanan Valley Volunteer Fire Department, Pennsylvania

Firefighter Cole was responding to the fire station after his fire department was dispatched to a wildland fire caused by arson. Firefighter Cole was driving his personal vehicle, a 2005 Subaru Impreza.

During his response, Firefighter Cole rounded a right-hand curve. He lost control of his vehicle and struck a car travelling in the opposite direction. The impact spun Firefighter Cole's vehicle counter-clockwise; his vehicle left the right side of the roadway and struck a tree.

Firefighter Cole was not wearing his seatbelt at the time of the crash. He was trapped between the passenger seat and the dash of his vehicle. Local emergency responders arrived and transported Firefighter Cole to the hospital, where he later died.

Excessive speed was cited in the law enforcement report on the crash as a contributing factor.

October 9, 2007–0855 hrs
Matthew Richard Will, Heavy Equipment Operator
Age 30–Wildland Full-Time
CAL-Fire, California

Heavy Equipment Operator Will was driving a heavy bulldozer as he and other firefighters fought the Colorado Fire near Monterey. During operations, the bulldozer rolled over and crushed Heavy Equipment Operator Will. He was airlifted to a regional hospital but did not survive his injuries.

October 25, 2007–0053 hrs
Robert Wayne Phillips, Firefighter
Age 64–Volunteer
Athelstane Volunteer Fire Department, Wisconsin

Firefighter Phillips and the members of his fire department responded to a chimney fire. Firefighter Phillips was operating the pump on one of the engine apparatus.

Firefighter Phillips collapsed due to an apparent heart attack. Other firefighters and EMS personnel came to his aid, and he was transported to a hospital. He was pronounced dead at 0053 hours on October 26, 2007.

continued on next page

October 27, 2007–1225 hrs
Ralph M. Cross, Assistant Fire Chief
Age 73–Volunteer
Charlevoix Township Fire Department, Michigan

Assistant Chief Cross was in the process of setting up a fire prevention information display at a local "big box" retail store. Assistant Chief Cross went out to the fire department vehicle that he was driving to retrieve boxes of handout material. When he returned to the store with a shopping cart full of materials, he suddenly collapsed.

Another firefighter who was participating in the event cared for Assistant Chief Cross until an ambulance arrived. Assistant Chief Cross was treated at the scene and transported to the hospital. He did not survive the effects of the heart attack.

October 29, 2007–1718 hrs
Scott A. Mumm, Lieutenant
Age 34–Paid-on-Call
Mendota Fire Department, Illinois

Lieutenant Mumm and the members of his fire department were dispatched to a fire in a coal hopper car at 1718 hours. The incident concluded at 2010 hours. Lieutenant Mumm was charged with filling tankers (tenders) at a fire hydrant during the firefighting effort.

At the conclusion of the incident, Lieutenant Mumm went home to rest and report to his full-time job at 0000 hours on October 30, 2007. At 2149 hours, firefighters were called to Lieutenant Mumm's residence for a medical emergency. Lieutenant Mumm was found not breathing and in cardiac arrest. He was pronounced dead at the hospital at 2245 hours. The cause of death was a heart attack.

November 5, 2007–0050 hrs
Jeremy W. Wach, Firefighter
Age 31–Volunteer
Wymore Fire & Rescue Department, Nebraska

Firefighter Wach responded along with other firefighters to a report of a fire in a residence. Firefighter Wach and two other firefighters advanced a handline into the structure to attack the fire. Shortly after they entered, the roof and ceiling collapsed and trapped Firefighter Wach in the debris.

The other two firefighters were able to withdraw from the structure, and firefighters quickly located Firefighter Wach. Due to his position and physical size, firefighters were unable to move Firefighter Wach. As fire conditions worsened, firefighters were ordered to abandon the building.

Firefighter Wach's death was confirmed at approximately 0200 hours, and his body was recovered from the building at approximately 0700 hours. His death was caused by traumatic positional asphyxiation.

The occupant of the residence had reported an electrical problem to the landlord two days before the fire. The fire likely had been burning in the attic for some time prior to the arrival of firefighters.

November 12, 2007–1610 hrs
Carl Stanley Engdahl, Fire Chief
Age 78–Volunteer
McPherson County Rural Fire Department #2, Kansas

Chief Engdahl and the members of his fire department responded to a kitchen fire at approximately 1549 hours. The incident was controlled, and the kitchen stove was removed from the structure.

As firefighters completed their duties on the scene, Chief Engdahl collapsed. Firefighters provided emergency medical care, and an ambulance was dispatched. Chief Engdahl was transported to the hospital where he later died as the result of a heart attack.

November 13, 2007–2241 hrs
Ronny Allen Bennett, Fire Chief
Age 39–Volunteer
Orcutt Fire Protection District, California

Chief Bennett and the members of his fire department were paged to respond to an emergency incident, but Chief Bennett did not arrive at the incident scene. Firefighters attempted to contact him numerous times without success.

The following day, he was found deceased at his residence. Firefighters found him with his radio and keys. He had apparently become ill and died as he began his response. The cause of death was listed as a heart attack.

continued on next page

November 21, 2007–1122 hrs
Michael J. Tluscik, Sr., Senior Firefighter
Age 48–Career
Kansas City Fire Department, Kansas

Firefighter Tluscik reported for duty and was assigned as the driver of Pumper 9. He completed his morning apparatus checks and station maintenance prior to 0800 hours. During the morning, Pumper 9 responded to a carbon monoxide alarm and a structure fire. During the response to the structure fire, Pumper 9 lost traction momentarily, and Firefighter Tluscik was able to gain control. He commented that this occurrence had shaken him. The pumper was cancelled prior to their arrival at the scene of the structure fire.

Firefighters returned to the fire station to eat their morning meal. Firefighter Tluscik sat on the couch and talked with other firefighters. Other firefighters noticed that Firefighter Tluscik was making snoring noises and that he was unresponsive.

Firefighters lowered Firefighter Tluscik to the floor and assessed him. CPR was initiated immediately, and paramedic-level EMS procedures were provided. Firefighter Tluscik was defibrillated at least three times prior to transport by ambulance to the hospital. He was pronounced dead at 1200 hours.

Firefighter Tluscik died of a cardiac condition known as mitral valve redundancy.

November 27, 2007–1057 hrs
John Jacob "JJ" Curry, Firefighter
Age 30–Career
Volusia County Fire Services, Florida

Firefighter Curry was a member of the Volusia County Wildland Fire Team. The team was practicing chain saw use at the fire department training center. During training, a pine tree fell and crushed Firefighter Curry. He was treated by firefighters at the scene but did not survive.

November 28, 2007–2330 hrs
Jon Mark Bingham, Assistant Fire Chief
Age 48–Volunteer
Geary Volunteer Fire Department, Oklahoma

Assistant Fire Chief Bingham responded with his department to a fatal apartment building fire late in the evening of November 18, 2007. According to reports, he stayed on scene over the next 18 hours until the recovery of the deceased was completed.

Chief Bingham returned to the station to put away equipment, finishing at approximately 1800 hours on November 29. Several hours after returning to his residence and going to bed, Assistant Fire Chief Bingham passed away from an apparent heart attack. He was pronounced dead at 0230 on November 30, 2007.

December 8, 2007–1741 hrs
Donald Wallis, Firefighter
Age 68–Volunteer
Forked River Volunteer Fire Company, New Jersey

On Saturday December 8, 2007, at 1700 hours, Fire Chief Kevin Flynn had the Forked River Fire Company (Station 60) dispatched to respond to their building for a holiday special detail that also promotes important community fire prevention, fire department recruiting, retention, and fundraising.

At 1741 hours, Firefighter Wallis suffered a massive heart attack while on duty performing his duties at the special detail. EMS and paramedics were requested and arrived on location at 1755 hours. Firefighter Wallis was transported to a local hospital where he was admitted into the critical care unit. Firefighter Wallis was not able to recover from his heart attack and died on December 16, 2007, at 0852 hours.

December 15, 2007–1112 hrs
Alphonse Vincent Germano, Jr., First Assistant Chief
Age 60–Volunteer
Derry Volunteer Fire Department, Pennsylvania

First Assistant Chief Germano returned from responding to an automatic fire alarm when the company was dispatched at 1112 hours to a residential fire. Chief Germano stayed behind with a standby crew and worked communications, and then assisted servicing apparatus when firefighters returned from the house fire. Shortly thereafter, fellow firefighters found Chief Germano in cardiac arrest.

Firefighters immediately initiated CPR, attached an AED that was provided by a responding police officer, and called for paramedics. Chief Germano was transported to the hospital, but died from the apparent heart attack.

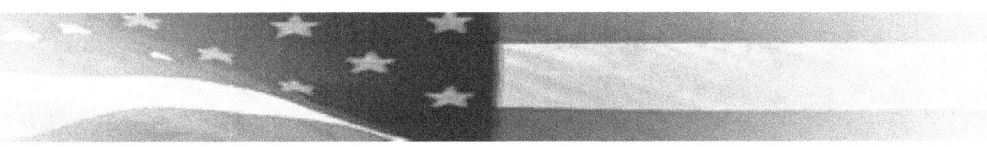

December 17, 2007–2300 hrs
Peter G. Neilson, Firefighter–Emergency Medical Responder
Age 74–Volunteer
Kenockee Township Fire Department, Michigan

At approximately 2300 hours, Firefighter Neilson responded with a crew to an emergency medical call. While on the scene, he was speaking with family members of the patient in an adjoining room when he sat down on a chair and suddenly passed out. Care was initiated immediately on scene, and he was transported to the local hospital where he succumbed to an apparent heart attack.

December 19, 2007–0330 hrs
Raymond Charles Simonis III, Firefighter
Age 48–Volunteer
Wissachickon Fire Company, Ambler, Pennsylvania

Firefighter Simonis responded as a part of a ladder company crew to a report of fire in a residence. Firefighters found a light haze and electrical smell in the house and determined that the source of the smell was a burned-out electrical motor. Firefighter Simonis had helped locate the source and was wearing full turnout gear.

Firefighters returned to the station and concluded the incident at approximately 1220 hours. Firefighter Simonis went home and later went to work.

Firefighter Simonis returned home from work in the early morning hours of December 18, 2007. He told his wife that he was having difficulty breathing. An ambulance was called and Firefighter Simonis was transported to the hospital. As Firefighter Simonis was transferred into the care of hospital staff, he became unconscious and died. The cause of death was a heart attack.

December 21, 2007–0900 hrs
Theresa Maria Lynn, Assistant Captain
Age 38–Volunteer
Luminary-Frostbite Volunteer Fire Department, Inc., Tennessee

Assistant Captain Lynn was responding in her personal vehicle to a reported vehicle crash with rollover. While responding in rain and fog conditions, she lost control of her vehicle in an "S-turn."

Her vehicle ran off the roadway and struck a tree. She succumbed to her injuries at the scene. The status of Assistant Captain Lynn's seatbelt at the time of the crash is unknown.

December 28, 2007–2353 hrs
Walter C. Fagan, Jr., Firefighter
Age 48–Volunteer
East Greenwich Township Fire/Rescue, New Jersey

Firefighter Fagan was responding to a report of a residential fire as a part of a fire company crew. The fire was found to be a good-intent call but the services of the fire department were not required.

Firefighter Fagan became ill while still in the apparatus. Firefighters provided assistance and called for an ambulance. Firefighter Fagan was transported to the hospital but died as the result of a heart attack.

Firefighter Fatality Prior to 2007

September 16, 2006–Time Unknown
Gary Curtis Cook, Firefighter
Age 49–Volunteer
Medina Lake Volunteer Fire Department, Inc., Texas

Firefighter Cook participated in ladder drills on the morning of September 16, 2006. Afternoon training included vehicle extrication activities using hand and power tools. Firefighter Cook responded from the training site to a wildland fire and participated in fire suppression activities.

At the conclusion of the wildland incident, Firefighter Cook returned his apparatus to service and departed for home at approximately 1700 hours. Firefighter Cook died later that day or early the next morning. He was discovered deceased on the morning of September 17, 2006. His death was caused by a heart attack.

Appendix B

FIREFIGHTER FATALITY INCLUSION CRITERIA–NATIONAL FIRE SERVICE ORGANIZATIONS

The National Fire Protection Association (NFPA), the National Fallen Firefighters Foundation (NFFF), the United States Fire Administration (USFA), and other organizations individually collect information on firefighter fatalities in the United States. Each organization uses a slightly different set of inclusion criteria, based at least, in part, on the purposes of the information collection for each organization and data consistency.

As a result of these differing inclusion criteria, statistics about firefighter fatalities may be provided by each organization that do not coincide with one another. This section will explain the inclusion criteria for each organization and provide information about these differences.

The USFA includes firefighters in this report who die while on duty, who become ill while on duty and later die, and firefighters who die within 24 hours of an emergency response or training, regardless of whether the firefighter complained of illness while on duty. The USFA counts firefighter deaths that occur in the 50 states, the District of Columbia, and United States Territories such as Puerto Rico and Guam. Detailed inclusion criteria for this report appear starting on page 90 of this report.

For 2007, the USFA reported 118 onduty firefighter fatalities.

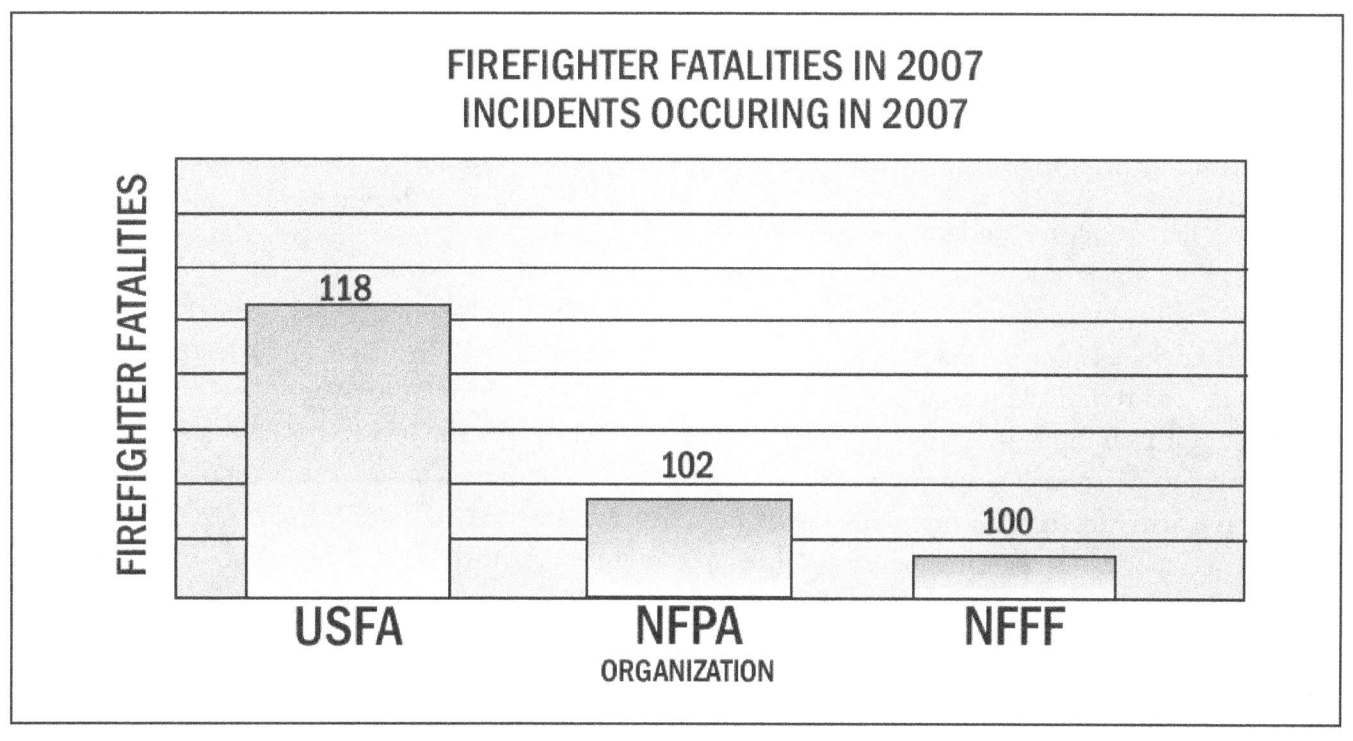

INCLUSION CRITERIA FOR NFPA'S ANNUAL FIREFIGHTER FATALITY STUDY

Introduction

Each year, the NFPA collects data on all firefighter fatalities in the United States that resulted from injuries or illnesses that occurred while the victims were on duty. The purpose of the study is to analyze trends in the types of illnesses and injuries resulting in death that occur while firefighters are on the job. This annual census of firefighter fatalities in its current format dates back to 1977. (Between 1974 and 1976, NFPA published a study of onduty firefighter fatalities that was not as comprehensive.)

WHAT IS A FIREFIGHTER?

For the purpose of the NFPA study, the term "firefighter" covers all uniformed members of organized fire departments, whether career, volunteer, combination, or contract; full-time public service officers acting as firefighters; State and Federal government fire service personnel; temporary fire suppression personnel operating under official auspices of one of the above; and privately employed firefighters including trained members of industrial or institutional fire brigades, whether full- or part-time.

Under this definition, the study includes, besides uniformed members of local career and volunteer fire departments, those seasonal and full-time employees of State and Federal agencies who have fire suppression responsibilities as part of their job description, prison inmates serving on firefighting crews, military personnel performing assigned fire suppression activities, civilian firefighters working at military installations, and members of industrial fire brigades. Impressed civilians also would be included if called on by the officer in charge of the incident to carry out specific duties. The NFPA study includes fatalities that occur in the 50 States and the District of Columbia.

WHAT DOES "ON DUTY" MEAN?

The term "on duty" refers to being at the scene of an alarm, whether a fire or nonfire incident; being en route while responding to or returning from an alarm; performing other assigned duties such as training, maintenance, public education, inspection, investigations, court testimony, or fundraising; and being on call, under orders, or on standby duty other than at home or at the individual's place of business. Fatalities that occur at a firefighter's home may be counted if the actions of the firefighter at the time of injury involved firefighting or rescue.

Onduty fatalities include any injury sustained in the line of duty that proves fatal, any illness that was incurred as a result of actions while on duty that proves fatal, and fatal mishaps involving nonemergency occupational hazards that occur while on duty. The types of injuries included in the first category are mainly those that occur at an incident scene, in training, or in accidents while responding to or returning from alarms. Illnesses (including heart attacks) are included when the exposure or onset of symptoms are tied to a specific incident of onduty activity. Those symptoms must have been in evidence while the victim was on duty for the fatality to be included in the study.

Fatal injuries and illnesses are included even in cases where death is delayed considerably. When the onset of the condition and the death occur in different years, the incident is counted in the year of the condition's onset. Medical documentation specifically tying the death to the specific injury is required for inclusion of these cases in the study.

CATEGORIES NOT INCLUDED IN THE STUDY

The NFPA study does not include members of fire department auxiliaries; nonuniformed employees of fire departments; emergency medical technicians who are not also firefighters; chaplains; or civilian dispatchers. The study also does not include suicides as onduty fatalities even when the suicide occurs on fire department property.

The NFPA recognizes that a comprehensive study of firefighter onduty fatalities would include chronic illnesses (such as cardiovascular disease and certain cancers) that prove fatal and that arose from occupational factors. In practice, there is as yet no mechanism for identifying onduty fatalities that are due to illnesses that develop over long periods of time. This creates an incomplete picture when comparing occupational illnesses to other factors as causes of firefighter deaths. This is recognized as a gap the size of which cannot be identified at this time because of the limitations in tracking the exposure of firefighters to toxic environments and substances and the potential long-term effects of such exposures.

2007 EXPERIENCE

In 2007, a total of 102 onduty firefighter deaths occurred in the United States, according to the NFPA inclusion criteria.

NATIONAL FALLEN FIREFIGHTERS FOUNDATION

In 1997, fire service leaders formulated new criteria to determine eligibility for inclusion on the National Fallen Firefighters Memorial. Line-of-duty deaths shall be determined by the following standards:

1. (a) Deaths of firefighters meeting the Department of Justice's Public Safety Officers' Benefits (PSOB) program guidelines, and those cases that appear to meet these guidelines whether or not PSOB staff has adjudicated the specific case prior to the annual National Fallen Firefighters Memorial Service; and

    (b) Deaths of firefighters from injuries, heart attacks, or illnesses documented to show a direct link to a specific emergency incident or department-mandated training activity.

2. While PSOB guidelines cover only public safety officers, the Foundation's criteria also include contract firefighters and firefighters employed by a private company, such as those in an industrial brigade, provided that the deaths meet the standards listed above.

3. Some specific cases will be excluded from consideration, such as deaths attributable to suicide, alcohol or substance abuse, or other gross abuses as specified in the PSOB guidelines.

The National Fallen Firefighters Memorial was built in 1981 in Emmitsburg, Maryland. The names listed there begin with those firefighters who died in the line-of-duty that year. The U.S. Congress created the National Fallen Firefighters Foundation to lead a nationwide effort to remember America's fallen firefighters. Since 1992, the tax-exempt, nonprofit Foundation has developed and expanded programs to honor our fallen fire heroes and assist their families and coworkers by providing them with resources to rebuild their lives. Since 1997, the Foundation has managed the National Memorial Service held each October to honor the firefighters who died in the line of duty the previous year.

At the October 2008 Memorial Weekend, the Foundation will be honoring 110 firefighters who died in the line of duty. Of those 110 being honored, 100 died in 2007 as the result of incidents that occurred in 2007, 1 firefighter died in 2007 as the result of an incident that occurred in a previous year, and 9 others died in previous years as the result of incidents that occurred in previous years. The following section is a listing of the firefighters who will be honored by the Foundation in October of 2008.

FIREFIGHTER DEATHS THAT OCCURRED IN 2007 AS THE RESULT OF AN INCIDENT THAT OCCURRED IN 2007

Sidney Alan Hall, Firefighter

Jeremy Christopher Adams, Fire Chief

Daryl W. Mutton, Captain

Kevin Charles Reed, Lieutenant

Shane Michael Daughetee, Firefighter

Anthony T. Catania, Fire Commissioner and Safety Officer

Craig Lawrence Dorsey II, Firefighter–EMT

Frederick Allen "Fred" Burroughs, Captain

Michael L. "Mike" Fox, Firefighter

Shane Todd King, Firefighter

John W. Broom-Smith, Jr., Firefighter

Jeremy Charles LaBella, Firefighter

Joseph Torkos, Fire Engineer Operator

Racheal Michelle Wilson, Firefighter–Paramedic Apprentice

Steven Eric Vanderpool, Jr., Firefighter

Jeffrey M. Murray, Firefighter

Theodore J. Abriel, Sr., Acting Lieutenant

Paul Tyler Reynolds, Sr., Firefighter–EMT

Eddy G. Ivers, Fire Chief

William F. Grant, Firefighter

Brandon Michael Whimple, Firefighter

Billy Harold Williams, Firefighter

Steve Olinik, Jr., Firefighter

Edgar Hamlin Scott, Fire Police Captain

Christopher Michael Jaros, Firefighter

Kyle Robert Wilson, Firefighter-Technician I

Billy Aaron Gafford, Jr., Captain–Paramedic

Joe Eddie Ivy, Captain

Vernon Robert "Bob" McKenzie, Firefighter

William D. Church, Sr., Fire Police Captain

Bruce Joseph Zumwalt, Firefighter

Peter Beebe-Lawson, Firefighter

Ottis Earl Stephenson, Jr., Firefighter

Brandon Lee Daley, Firefighter

Jeffrey Jeans, Firefighter–EMT

Edward Levirgil Andrews IV, Fire Captain

Joseph E. Piazzi, Deputy Chief and Fire Police Captain

John Francis Keane, Captain

Dennis Cheshire, Firefighter

Bradley William "Wally" Green, Firefighter

Henry Lawrence "Hank" Pitts, Jr., Lieutenant

Felix Maurice Roberts, Firefighter II

David A. Middleton, Firefighter

Mark Stevens Carter, Engineer–Paramedic

David Allan Rufer, Firefighter

William H. "Billy" Hutchinson, Captain

Theodore Michael Benke, Captain

Louis Mark Mulkey, Captain

Mark Wesley Kelsey, Engineer-Acting Captain

Bradford Rodney "Brad" Baity, Engineer

Melvin Edward Champaign, Firefighter

Michael Jonathon Alan French, Assistant Engineer

Brandon Kenyon Thompson, Firefighter

James "Earl" Allen Drayton, Firefighter

Daniel F. Pujdak, Firefighter

Timothy Lavern Sanborn, Firefighter

Samuel W. Downing, Captain

Dennise Marie Leslie, Firefighter

James J. McRae III, Firefighter–Technician

Stephen R. Dembski, Firefighter

Matthew Charles Burton, Captain

Scott Peter Desmond, Fire Engineer

Dennis Luster Davis, Pilot

Jon Charles Trainer, Firefighter

James Glenn "Shib" Miller, Firefighter

Nemeth Fitzhugh Sanders, Forest Fire Equipment Operator

Kevin Glenn Williams, Captain

Austin Hague Cheek, Firefighter

Gerald David "Jerry" Donley, Jr., Deputy Fire Chief

Paul Darrin Baker, Lieutenant

Todd Whitney Hage, Firefighter

Anthony Philip Cox, Captain

Glenn Williams Miller, Probationary Firefighter

Michael Paul Stephen Heuer, Fire Chief

Robert C. Beddia, Firefighter

Joseph Graffagnino, Firefighter

Jeffrey Dale Swartz, Firefighter

Jared W. Zimmerly, Firefighter

Michael Dean Stanfield, Captain

Leonard R. Bailey, Jr., Fire Chief

Jerry Wayne Stucker, Firefighter–EMT

Kelly L. Page, Firefighter

William D. "Billy" McDaniels, Sr., Fire Police Captain

Bryon Wayne Johnson, Lieutenant

John E. "Dawg" Lietzke, Firefighter

Mike D. Reagan, Jr., Firefighter

Adam Edward Cole, Firefighter

Matthew Richard Will, Heavy Fire Equipment Operator

Robert Wayne Phillips, Firefighter

Scott A. Mumm, Lieutenant

Jeremy W. Wach, Firefighter

Carl Stanley Engdahl, Fire Chief

Michael J. Tluscik, Sr., Senior Firefighter

John Jacob "JJ" Curry, Firefighter

Jon Mark Bingham, Assistant Fire Chief

Alphonse Vincent Germano, Jr., First Assistant Chief

Peter G. Neilson, Firefighter–Medical First Responder

Raymond Charles Simonis III, Firefighter

Theresa Maria Lynn, Assistant Captain

Walter C. Fagan, Jr., Firefighter

FIREFIGHTER DEATH THAT OCCURRED IN 2007 FROM AN INCIDENT IN A PREVIOUS YEAR

Stephen Harbison, Firefighter–Paramedic

FIREFIGHTER DEATHS THAT OCCURRED IN PREVIOUS YEARS

Paul R. Brady, Firefighter

Gary Vernon Cook, Captain

Chelsea Lyn Garvin, Firefighter

Darrell Keith Michael, President

William A. Riley, Firefighter

Benjamin Craig Rouchon, Firefighter

David Wayne Stautamoyer, Assistant Fire Chief

John A. Stura, Firefighter

Gregory Harold Vieth, Lieutenant

Photo by Mark Whitney, U.S. Fire Administration

94

www.ingramcontent.com/pod-product-compliance
Lightning Source LLC
Chambersburg PA
CBHW081140170526
45165CB00008B/2744